Lecture Notes in Computer Science 2048

Edited by G. Goos, J. Hartmanis and J. van Leeuwen

Springer

*Berlin
Heidelberg
New York
Barcelona
Hong Kong
London
Milan
Paris
Singapore
Tokyo*

Josef Pauli

Learning-Based Robot Vision

Principles and Applications

 Springer

Series Editors

Gerhard Goos, Karlsruhe University, Germany
Juris Hartmanis, Cornell University, NY, USA
Jan van Leeuwen, Utrecht University, The Netherlands

Author

Josef Pauli
Christian-Albrecht Universiät zu Kiel
Institut für Informatik und Praktische Mathematik
Preusserstr. 1-9, 24105 Kiel, Germany
E-mail: jpa@ks.informatik.uni-kiel.de

Cataloging-in-Publication Data applied for

Die Deutsche Bibliothek - CIP-Einheitsaufnahme

Pauli, Josef:
Learning-based robot vision : principles and applications / Josef
Pauli. - Berlin ; Heidelberg ; New York ; Barcelona ; Hong Kong ;
London ; Milan ; Paris ; Singapore ; Tokyo : Springer, 2001
  (Lecture notes in computer science ; 2048)
  ISBN 3-540-42108-4

CR Subject Classification (1998): I.4, I.2.9-11, I.2.6

ISSN 0302-9743
ISBN 3-540-42108-4 Springer-Verlag Berlin Heidelberg New York

Springer-Verlag Berlin Heidelberg New York
a member of BertelsmannSpringer Science+Business Media GmbH

http://www.springer.de

© Springer-Verlag Berlin Heidelberg 2001
Printed in Germany

Typesetting: Camera-ready by author, data conversion by Boller Mediendesign
Printed on acid-free paper   SPIN 10781470   06/3142   5 4 3 2 1 0

Preface

Industrial robots carry out simple tasks in customized environments for which it is typical that nearly all effector movements can be planned during an off-line phase. A continual control based on sensory feedback is at most necessary at effector positions near target locations utilizing torque or haptic sensors. It is desirable to develop new-generation robots showing higher degrees of autonomy for solving high-level deliberate tasks in natural and dynamic environments. Obviously, camera-equipped robot systems, which take and process images and make use of the visual data, can solve more sophisticated robotic tasks. The development of a (semi-) autonomous camera-equipped robot must be grounded on an infrastructure, based on which the system can acquire and/or adapt task-relevant competences autonomously. This infrastructure consists of technical equipment to support the presentation of real world training samples, various learning mechanisms for automatically acquiring function approximations, and testing methods for evaluating the quality of the learned functions. Accordingly, to develop autonomous camera-equipped robot systems one must first demonstrate relevant objects, critical situations, and purposive situation-action pairs in an experimental phase prior to the application phase. Secondly, the learning mechanisms are responsible for acquiring image operators and mechanisms of visual feedback control based on supervised experiences in the task-relevant, real environment.

This paradigm of learning-based development leads to the concepts of compatibilities and manifolds. Compatibilities are general constraints on the process of image formation which hold more or less under task-relevant or accidental variations of the imaging conditions. Based on learned degrees of compatibilities, one can choose those image operators together with parametrizations, which are expected to be most adequate for treating the underlying task. On the other hand, significant variations of image features are represented as manifolds. They may originate from changes in the spatial relation among robot effectors, cameras, and environmental objects. Learned manifolds are the basis for acquiring image operators for task-relevant object or situation recognition. The image operators are constituents of task-specific, behavioral modules which integrate deliberate strategies and visual feedback control. The guiding line for system development is that the resulting behaviors should meet requirements such as task-relevance, robustness, flexibility, time limitation, *etc.* simultaneously. All principles to be presented in the work are based on real scenes of man-made objects and a multi-component robot system consisting of robot arm, head, and vehicle. A high-level application is presented that includes sub-tasks such as localizing, approaching, grasping, and carrying objects.

Acknowledgements

Since 1993 I have been a member of the Kognitive Systeme Gruppe in Kiel. I am most grateful to G. Sommer, the head of this group, for the continual advice and support. I have learned so much from his wide spectrum of scientific experience.

A former version of this book has been submitted and accepted as habilitation thesis at the Institut für Informatik und Praktische Mathematik, Technische Fakultät of the Christian-Albrechts-Universität, in Kiel. I'm very grateful to the six persons who have been responsible for assessing the work. These are: V. Hlavac from the Technical University of Prague in the Czech Republic, R. Klette from the University of Auckland in New Zealand, C.-E. Liedtke from the Universität Hannover, G. Sommer and A. Srivastav from the Christian-Albrechts-Universität Kiel, and F. Wahl from the Technische Universität Braunschweig. Deepest thanks also for the great interest in my work.

I appreciate the discussions with former and present colleagues J. Bruske, T. Bülow, K. Daniilidis, M. Felsberg, M. Hansen, N. Krüger, V. Krüger, U. Mahlmeister, C. Perwass, B. Rosenhahn, W. Yu, and my students M. Benkwitz, A. Bunten, S. Kunze, F. Lempelius, M. Päschke, A. Schmidt, W. Timm, and J. Tröster.

Technical support was provided by A. Bunten, G. Diesner, and H. Schmidt. My thanks to private individuals have been expressed personally.

March 2001   *Josef Pauli*

Contents

1. Introduction

The first chapter presents an extensive introduction to the book by starting with the motivation. Next, the *Robot Vision* paradigm is characterized
and confronted with the field of *Computer Vision*. Robot Vision is the indisputable kernel of *Autonomous Camera-Equipped Robot Systems*. For the development of such new-generation robot systems the important role of visual
demonstration and learning is explained. The final section gives an overview
to the chapters of the book.

1.1 Need for New-Generation Robot Systems

We briefly describe present state and problems of robotics, give an outlook
on trends of research and development, and summarize the specific novelty
contributed in this book.

Present State of Robotics

Industrial robots carry out recurring simple tasks in a fast, accurate and
reliable manner. This is typically the case in applications of series production.
The environment is customized in relation to a fixed location and volume
occupied by the robot and/or the robot is built such that certain spatial
relations with a fixed environment are kept. Task-relevant effector trajectories
must be planned perfectly during an offline phase and unexpected events
must not occur during the subsequent online phase. Close-range sensors are
utilized (if at all) for a careful control of the effectors at the target positions.
Generally, sophisticated perception techniques and learning mechanisms, *e.g.*
involving Computer Vision and Neural Networks, are unnecessary due to
customized relations between robot and environment.

   In the nineteen eighties and nineties impressive progress has been achieved
in supporting the development and programming of industrial robots.

- CAD (Computer Aided Design) tools are used for convenient and rapid
  designing of the hardware of robot components, for example, shape and
  size of manipulator links, degrees-of-freedom of manipulator joints, *etc.*

J. Pauli: Learning-Based Robot Vision, LNCS 2048, pp. 1-24, 2001.
© Springer-Verlag Berlin Heidelberg 2001

- Application-specific signal processors are responsible for the control of the motors of the joints and thus cope with the dynamics of articulated robots. By solving the inverse kinematics the effectors can be positioned up to sub-millimeter accuracy, and the accuracy does not degrade even for high frequencies of repetition.
- High-level robot programming languages are available to develop specific programs for executing certain effector trajectories. There are several methodologies to automate the work of programming.
- Teach-in techniques rely on a demonstration of an effector trajectory, which is executed using a control panel, and the course of effector coordinates is memorized and transformed into a sequence of program-steps.
- Automatic planning systems are available which generate robot programs for assembly or disassembly tasks, *e.g.* sequences of movement steps of the effector for assembling complex objects out of components. These systems assume that initial state and desired state of the task are known accurately.
- Appropriate control mechanisms are applicable to fine-control the effectors at the target locations. It is based on sensory feedback from close-range sensors, *e.g.* torque or haptic sensors.

This development kit consisting of tools, techniques and mechanisms is widely available for industrial robots. Despite of that, there are serious limitations concerning the possible application areas of industrial robots. In the following, problems and requirements in robotics are summarized, which serves as a motivation for the need for advanced robot systems. The mentioned development kit will be a part of a more extensive infrastructure which is necessary for the creation and application of new-generation robots.

Problems and Requirements in Robotics

The lack of a camera subsystem and of a conception for making extensive use of environmental data is a source of many limitations in industrial robots.

- In the long term an insidious wear of robot components will influence the manufacturing process in an unfavourable manner which may lead to unusable products. For exceptional cases the robot effector may even damage certain environmental components of the manufacturing plant.
- Exceptional, non-deterministic incidents with the robot or in the environment, *e.g.* break of the effector or dislocated object arrangement, need to be recognized automatically in order to stop the robot and/or adapt the planned actions.
- In series production the variance of geometric attributes must be tight respectively from object to object in the succession, *e.g.* nearly constant size, position, and orientation. Applications of frequently changing situations, *e.g.* due to the object variety, can not be treated by industrial robots.
- The mentioned limitations will cause additional costs which contribute to the overall manufacturing expenses. These costs can be traced back to

the production of unusable products, the loss of production due to offline adaptation, the damage of robot equipment, *etc.*

The main methodology to overcome these problems is to perceive the environment continually and make use of the reconstructed spatial relations between robot effector and target objects. In addition to the close-range sensors one substantially needs long-range perception devices such as video, laser, infrared, and ultrasonic cameras. The long-range characteristic of cameras is appreciated for early measuring effector-object relations in order to adapt the effector movement timely (if needed). The specific limitations and constraints, which are inherent in the different perception devices, can be compensated by a fusion of the different image modalities. Furthermore, it is advantageous to utilize steerable cameras which provide the opportunity to control external and internal degrees-of-freedom such as pan, tilt, vergence, apperture, focus, and zoom. Image analysis is the basic means for the primary goal of reconstructing the effector-object relations, but also the prerequisite for the secondary goal of information fusion and camera control. To be really useful, the image analysis system must extract purposive information in the available slice of time.

The application of camera-equipped robots (in contrast to blind industrial robots) could lead to damage prevention, flexibility increase, cost reduction, *etc.* However, the extraction of relevant image information and the construction of adequate image-motor mappings for robot control causes tremendous difficulties. Generally, it is hard if not impossible to proof the correctness of reconstructed scene information and the goal-orientedness of image-motor mappings. This is the reason why the development and application of camera-equipped robots is restricted to (practically oriented) research institutes. So far, industries still avoid their application, apart from some exceptional cases. The components and dynamics of more or less natural environments are too complex and therefore imponderabilities will occur which can not be considered in advance. More concretely, quite often the procedures to be programmed for image analysis and robot control are inadequate, non-stable, inflexible, and inefficient. Consequently, the development and application of new-generation robots must be grounded on a learning paradigm. For supporting the development of autonomous camera-equipped robot systems, the nascent robot system must be embedded in an infrastructure, based on which the system can learn task-relevant image operators and image-motor mappings. In addition to that, the robot system must be willing to make life-long experience and adapt the behaviors for new environments.

New Application Areas for Camera-Equipped Robot Systems

In our opinion, leading-edge robotics institutes agree with the presented catalog of problems and requirements. Pursuing the mission to strive for autonomous robots each institute individually focuses on and treats some of the problems in detail. As a result, new-generation robot systems are being developed, which can be regarded as exciting prototype solutions. Frequently, the robots show increased robustness and flexibility for tasks which need to be solved in non-customized environments. The robots behaviors are purposive despite of large variations of environmental situations or even in cases of exceptional, non-deterministic incidents. Consequently, by applying new-generation robots to classical tasks (up to now performed by industrial robots), it should be possible to relax the customizing of the environment. For example, the manufacturing process can be organized more flexible with the purpose of increasing the product variety.[1]

Beyond manufacturing plants, which are typical environments of industrial robots, the camera-equipped robots should be able to work purposive in completely different (more natural) environments and carrying out new categories of tasks. Examples of such tasks include supporting disabled persons at home, cleaning rooms in office buildings, doing work in hazardous environments, automatic modeling of real objects or scenes, *etc*. These tasks have in common that objects or scenes must be detected in the images and reconstructed with greater or lesser degree of detail. For this purpose the agility of a camera-equipped robot is exploited in order to take environmental images under controlled camera motion. The advantages are manifold, for example, take a degenerate view to simplify specific inspection tasks, take various images under several poses to support and verify object recognition, take an image sequence under continual view variation for complete object reconstruction.

The previous discussion presented an idea of the wide spectrum of potential application areas for camera-equipped robot systems. Unfortunately, despite of encouraging successes achieved by robotics institutes, there are still tremendous difficulties in creating really usable camera-equipped robot systems. In practical applications these robot systems are lacking correct and goal-oriented image-motor mappings. This finding can be traced back to the lack of correctness of image processing, feature extraction, and reconstructed scene information. We have to have new conceptions for the development and evaluation of image analysis methods and image-motor mappings.

Contribution and Novelty of This Book

This work introduces a practical methodology for developing autonomous camera-equipped robot systems which are intended to solve high-level, deliberate tasks. The development is grounded on an infrastructure, based on

---

[1] The german engineer newspaper VDI-Nachrichten reported in the issue of February 18, 2000, that BMW intends to make investments of 30 billion Deutsche Marks for the development of highly flexible manufacturing plants. A major aim is to develop and apply more flexible robots which should be able to simultaneously build different versions of BMW cars on each manufacturing plant. The spectrum of car variety must not be limited by unflexible manufacturing plants, but should only depend on specific demands on the market.

which the system can learn competences by interaction with the real task-relevant world. The infrastructure consists of technical equipment to support the demonstration of real world training samples, various learning mechanisms for automatically acquiring function approximations, and testing methods for evaluating the quality of the learned functions. Accordingly, the application phase must be preceded by an experimental phase in order to construct image operators and servoing procedures, on which the task-solving process mainly relies. Visual demonstration and neural learning is the backbone for acquiring the situated competences in the real environment.

This paradigm of *learning-based development* distinguishes between two learnable categories: compatibilities and manifolds. Compatibilities are general constraints on the process of image formation, which do hold to a certain degree. Based on learned degrees of compatibilities, one can choose those image operators together with parametrizations, which are expected to be most adequate for treating the underlying task. On the other hand, significant variations of image features are represented as manifolds. They may originate from changes in the spatial relation among robot effectors, cameras, and environmental objects. Learned manifolds are the basis for acquiring image operators for task-relevant object or situation recognition. The image operators are constituents of task-specific, behavioral modules which integrate deliberate strategies and visual feedback control. As a summary, useful functions for image processing and robot control can be developed on the basis of learned compatibilities and manifolds.

The practicality of this development methodology has been verified in several applications. In the book, we present a structured application that includes high-level sub-tasks such as localizing, approaching, grasping, and carrying objects.

1.2 Paradigms of Computer Vision (CV) and Robot Vision (RV)

The section cites well-known definitions of Computer Vision and characterizes the new methodology of Robot Vision.

1.2.1 Characterization of Computer Vision

Almost 20 years ago, Ballard and Brown introduced a definition for the term *Computer Vision* which was commonly accepted until present time [11].

Definition 1.1 (Computer Vision, according to Ballard) *Computer Vision is the construction of explicit, meaningful descriptions of physical objects from images. Image processing, which studies image-to-image transformations, is the basis for explicit description building. The challenge of Computer Vision is one of explicitness. Explicit descriptions are a prerequisite for recognizing, manipulating, and thinking about objects.*

In the nineteen eighties and early nineties the research on *Artificial Intelligence* influenced the Computer Vision community [177]. According to the principle of Artificial Intelligence, both common sense and application-specific knowledge are represented explicitly, and reasoning mechanisms are applied (*e.g.* based on *predicate calculus*) to obtain a *problem solver* for a specific application area [119]. According to this, explicitness is essential in both Artificial Intelligence and Computer Vision. This coherence inspired Haralick and Shapiro to a definition of Computer Vision which uses typical terms of Artificial Intelligence [73].

Definition 1.2 (Computer Vision, according to Haralick) *Computer Vision is the combination of image processing, pattern recognition, and artificial intelligence technologies which focuses on the computer analysis of one or more images, taken with a singleband/multiband sensor, or taken in time sequence. The analysis recognizes and locates the position and orientation, and provides a sufficiently detailed symbolic description or recognition of those imaged objects deemed to be of interest in the three-dimensional environment. The Computer Vision process often uses geometric modeling and complex knowledge representations in an expectation- or model-based matching or searching methodology. The searching can include bottom-up, top-down, blackboard, hierarchical, and heterarchical control strategies.*

Main Issues of Computer Vision

The latter definition proposes to use Artificial Intelligence technologies for solving problems of representation and reasoning. The interesting objects must be extracted from the image leading to a description of the 2D image situation. Based on that, the 3D world situation must be derived. At least four main issues are left open and have to be treated in any Computer Vision system.

1. Which types of representation for 3D world situations are appropriate ?
2. Where do the models for detection of 2D image situations originate ?
3. Which reasoning or matching techniques are appropriate for detection tasks ?
4. How should the gap between 2D image and 3D world situations be bridged ?

Non-realistic Desires in Computer Vision

This paradigm of Computer Vision resembles the enthusiastic work in the nineteen sixties on developing a *General Problem Solver* [118]. Nowadays, the efforts for a General Problem Solver appear hopeless and ridiculous, and it is similarly ridiculous to strive for a *General Vision System*, which is supposed to solve any specific vision task [2]. Taking the four main issues of Computer Vision into account, a general system would have to include the following four characteristics.

1. A unifying representation framework for dealing with various representations of signals and symbols.
2. Common modeling tools for acquiring models, *e.g.* for reconstruction from images or for generation of CAD data.
3. General reasoning techniques (*e.g.* in fuzzy logic) for extracting relevant image structures, or general matching procedures for recognizing image structures.
4. General imaging theories to model the mapping from 3D world into 2D images (executed by the cameras).

Continuing with the train of thought, a General Vision System would have to be designed as a shell. This is quite similar to *Expert System Shells* which include general facilities of knowledge representation and reasoning. Various categories of knowledge, ranging from specific scene/task knowledge to general knowledge about the use of image processing libraries, are supposed to be acquired and filled into the shell on demand. Crevier and Lepage present an extensive survey of knowledge-based image understanding systems [43], however, they concede that *"genuine general-purpose image processing shells do not yet exist."* In summary, representation frameworks, modeling tools, reasoning and matching techniques, and imaging theories are not available in the required generality.

Favouring Robot Vision in Opposition to Computer Vision

The statement of this book is that the required generality can never be reached, and that degradations in generality are acceptable in practical systems. However, current Computer Vision systems (in industrial use) only work well for specific scenes under specific imaging conditions. Furthermore, this specificity has also influenced the design process, and, consequently, there is no chance to adapt a classical system to different scenes.

New design principles for more general and flexible systems are necessary in order to overcome to a certain extent the large gap between general desire and specific reality.

These principles can be summarized briefly by *animated attention, purposive perception, visual demonstration, compatible perception, biased learning,* and *feedback analysis.* The following discussion will reveal that all principles are closely connected with each other. The succinct term *Robot Vision* is used for systems which take these principles into account.[2]

1.2.2 Characterization of Robot Vision

Animated Vision by Attention Control

It is assumed that most of the three-dimensional vision-related applications must be treated by analyzing images at different viewing angles and/or distances [12, 1]. Through exploratory controlled camera movement the system gathers information incrementally, *i.e.* the environment serves as external memory from which to read on demand. This paradigm of animated vision also includes mechanisms of selective attention and space-variant sensing [40]. Generally, a two-part strategy is involved consisting of attention control and detailed treatment of the most interesting places [145, 181]. This approach is a compromise for the trade-off between effort of computations and sensing at high resolution.

Purposive Visual Information

Only that information of the environmental world must be extracted from the images which is relevant for the vision task. The modality of that information can be of quantitative or qualitative nature [4]. In various phases of a Robot Vision task presumably different modalities of information are useful, *e.g.* color information for tracking robot fingers, and geometric information for grasping objects. The minimalism principle emphasizes to solve the task by using features as basic as possible [87], *i.e.* avoiding time-consuming, erroneous data abstraction and high-level image representation.

Symbol Grounding by Visual Demonstration

Models, which represent target situations, will only prove useful if they are acquired in the same way, or under the same circumstances, as when the system perceives the scene in real application [75]. It is important to have a close relation between physically grounded task specifications and the appearance of actual situations [116]. Furthermore, it is easier for a person to specify target situations by demonstrating examples instead of describing visual tasks symbolically. Therefore, visual demonstration overcomes the necessity of determining quantitative theories of image formation.

Perception Compatibility (Geometry/Photometry)

In the imaging process, certain compatibilities hold between the (global) geometric shape of the object surface and the (local) gray value structure in the photometric image [108]. However, there is no one-to-one correspondence

---

[2] The adequacy will become obvious later on.

between surface discontinuities and extracted gray value edges, *e.g.* due to texture, uniform surface color, or lighting conditions. Consequently, qualitative compatibilities must be exploited, which are generally valid for certain classes of regular objects and certain types of camera objectives, in order to bridge the global-to-local gap of representation.

Biased Learning of Signal Transformation

The signal coming from the imaging process must be transformed into 2D or 3D features, whose meaning depends on the task at hand, *e.g.* serving as motor signal for robot control, or serving as symbolic description for a user. This transformation must be learned on the basis of samples, as there is no theory for determining it *a priori*. Each signal is regarded as a point in an extremely high-dimensional space, and only a very small fraction will be considered by the samples of the transformation [120]. Attention control, visual demonstration, and geometry/photometry compatibilities are taken as bias for determining the transformation, which is restricted to a relevant signal sub-space.

Feedback-Based Autonomous Image Analysis

The analysis algorithms used for signal transformation require the setting or adjustment of parameters [101]. A feedback mechanism is needed to reach autonomy instead of adjusting the parameters interactively [180]. A cyclic process of quality assessment, parameter adjustment, and repeated application of the algorithm can serve as backbone of an automated system [126].

For the vast majority of vision-related tasks only Robot Vision systems can provide pragmatic solutions. The possibility of camera control and selective attention should be exploited for resolving ambiguous situations and for completing task-relevant information. The successful execution of the visual task is critically based on autonomous learning from visual demonstration. The online adaptation of visual procedures takes possible deviations between learned and actual aspects into account. Learning and adaptation are biased under general compatibilities between geometry and photometry of image formation, which are assumed to hold for a category of similar tasks and a category of similar camera objectives.

> General representation frameworks, reasoning techniques, and imaging theories are no longer needed, rather, task-related representations, operators, and calibrations are learned and adapted on demand.

The next Section 1.3 will demonstrate that these principles of Robot Vision are in consensus with new approaches to designing autonomous robot systems.

1.3 Robot Systems versus Autonomous Robot Systems

Robots work in environments which are more or less customized to the dimension and the needs of the robot.

1.3.1 Characterization of a Robot System

Definition 1.3 (Robot System) *A robot system is a mechanical device which can be programmed to move in the environment and handle objects or tools. The hardware consists essentially of an actuator system and a computer system. The actuator system is the mobile and/or agile body which consists of the effector component (exterior of the robot body) and the drive component (interior of the robot body). The effectors physically interact with the environment by steering the motors of the drive. Examples for effectors are the wheels of a mobile robot (robot vehicle) or the gripper of a manipulation robot (manipulator, robot arm). The computer system is composed of general and/or special purpose processors, several kinds of storage, etc., together with a power unit. The software consists of an interpreter for transforming high-level language constructs into an executable form and procedures for solving the inverse kinematics and sending steering signals to the drive system.*

Advanced robot systems are under development which will be equipped with a sensor or camera system for perceiving the environmental scene. Based on perception, the sensor or camera system must impart to the robot an impression of the situation wherein it is working, and thus the robot can take appropriate actions for more flexibly solving a task. The usefulness of the human visual system gives rise to develop robots equipped with video cameras. The video cameras of an advanced robot may or may not be a part of the actuator system.

In camera-equipped systems the robots can be used for two alternative purposes leading to a *robot-supported vision system (robot-for-vision tasks)* or to a *vision-supported robot system (vision-for-robot tasks)*. In the first case, a purposive camera control is the primary goal. For the inspection of objects, factories, or processes, the cameras must be agile for taking appropriate images. A separate actuator system, *i.e.* a so-called *robot head*, is responsible for the control of external and/or internal camera parameters. In the second case, cameras are fastened on a stable tripod (*e.g. eye-off-hand system*) or fastened on an actuator system (*e.g. eye-on-hand system*), and the images are a source of information for the primary goal of executing robot tasks autonomously. For example, a manipulator may handle a tool on the basis of images taken by an eye-off-hand or an eye-on-hand system. In both cases, a dynamic relationship between camera and scene is characteristic, *e.g.* inspecting situations with active camera robots, or handling tools with vision-based manipulator robots. For more complicated applications the cameras must be separately agile in addition to the manipulator robot, *i.e.* having a robot of its own just for the control of the cameras. For those advanced arrangements, the distinction between robot-supported vision system and vision-supported robot system no longer makes sense, as both types are fused.

The most significant issue in current research on advanced robot systems is to develop an *infrastructure*, based on which a robot system can learn and adapt task-relevant competences autonomously. In the early nineteen nineties, Brooks made clear in a series of papers that the development of autonomous robots must be based on completely new principles [26, 27, 28]. Most importantly, autonomous robots can not emerge by simply combining results from research on Artificial Intelligence and Computer Vision. Research in both fields concentrated on reconstructing symbolic models and reasoning about abstract models, which was quite often irrelevant due to unrealistic assumptions. Instead of that, an intelligent system must interface directly to the real world through perception and action. This challenge can be handled by considering four basic characteristics that are tightly connected with each other, *i.e. situatedness*, *corporeality*, *emergence*, and *competence*. Autonomous robots must be designed and organized into task-solving behaviors, taking the four basic characteristics into account.[3]

1.3.2 Characterization of an Autonomous Robot System

Situatedness

The autonomous robot system solves the tasks in the total complexity of concrete situations of the environmental world. The task-solving process is based on situation descriptions, which must be acquired continually using sensors and/or cameras. Proprioceptive and exteroceptive features of a situation description are established, which must be adequate and relevant for solving the specific robot task at hand. Proprioceptive features describe the internal state of the robot, *e.g.* the coordinates of the tool center point, which can be changed by the inherent degrees of freedom. Exteroceptive features describe aspects in the environmental world and, especially, the relationship between robot and environment, *e.g.* the distance between robot hand and target object. The characteristic of a specific robot task is directly correlated with a certain type of situation description. For example, for robotic object grasping the exteroceptive features describe the geometric relation between the shape of the object and the shape of the grasping fingers. However, for robotic object inspection another type of situation description is relevant, *e.g.* the silhouette contour of the object. Based on the appropriate type of situation description, the autonomous robot system must continually interpret and evaluate the concrete situations correctly.

---

[3] In contrast to Brooks [27], we prefer the term *corporeality* instead of *embodiment* and *competence* instead of *intelligence*, both replacements seem to be more appropriate (see also Sommer [161]).

Corporeality

The camera-equipped robot system experiences the world under physical constraints of the robot and optical constraints of the camera. These robot and camera characteristics affect the task-solving process crucially. Robots and cameras are themselves part of the scene, and therefore in a situation description the proprioceptive features are correlated with the exteroceptive features. For example, if a camera is fastened on a robot hand with the optical axis pointing directly through parallel jaw fingers, then the closing of the fingers is reflected both in the proprioceptive and exteroceptive part of the situation description. The specific type of the camera system must be chosen to be favourable for solving the relevant kind of tasks. The camera system can be static, or fastened on a manipulator, or separate and agile using a robot head, or various combinations of these arrangements. The perception of the environmental world is determined by the type of camera objectives, *e.g.* the focal length influences the depicted size of an object, the field of view, and possible image distortions. Therefore, useful classes of 2D situations in the images can only be acquired based on actually experienced 3D situations. The specific characteristic of robot and camera must be taken into account directly without abstract modeling. A purposive robotic behavior is characterized by situatedness and is biased by the corporeality of the robotic equipment.

Competence

An autonomous robot system must show competent behaviors when working on tasks in the real environment. Both expected and unforeseen environmental situations may occur upon which the task-solving process has to react appropriately. The source of competence originates in signal transformations, mainly including feature extraction, evaluation of situations and construction of mappings between situations and actions. Environmental situations may represent views of scene objects, relations between scene objects, or relations between robot body and scene objects. The specific type of situation description is determined under the criterion of task-relevance including a minimalism principle. That is, from the images only the minimum amount of quantitative or qualitative information should be extracted which is absolutely necessary for solving the specific task. In order to reach sufficient levels of competence, the procedures for minimalistic feature extraction and situation evaluation can hardly be programmed, but have to be learned on the basis of demonstrations in an offline phase. Additionally, purposive situation-action pairs have to be learned which serve as ingredients of visual servoing mechanisms. A servoing procedure must catch the actual dynamics of the relationship between effector movements and changing environmental situations, *i.e.* actual dynamics between proprioceptive and exteroceptive features. The purpose is to reach a state of equilibrium between robot and environment, without having exact models from environment and camera.

Emergence

The competences of an autonomous robot system must emerge from the system's interactions with the world. The learned constructs and servoing mechanisms can be regarded as backbone for enabling these interactions. In the online phase the learned constructs work well for expected situations leading to basic competences in known environments. However, degradations occur in environments of deviating or unforeseen situations. The confrontation with actual situations causes the system to revisit and maybe adapt certain ingredients of the system on demand. Consequently, more robust competences are being developed by further training and refining the system on the job. Furthermore, competences at higher levels of complexity can be developed. The repeated application of servoing cycles give rise to group perception-action pairs more compactly which will lead to macro actions. Additionally, based on exploration strategies applied in unknown environments, the system can collect data with the purpose of building a map, and based on that, constructing more purposive situation-action pairs. By considering concepts as mentioned before, high level competences will emerge from low level competences which in turn are based on learned constructs and servoing mechanisms. Common sense knowledge about methodologies of situation recognition or strategies of scene exploration should be made available to the system. This knowledge reduces the need to let learn and/or emerge everything. However, before taking this knowledge for granted, the validity must be tested in advance under the actual conditions.

Behavioral Organization

Each behavior is based on an activity producing subsystem, featuring sensing, processing, and acting capabilities. The organization of behaviors begins on the bottom level with very simple but complete subsystems, and follows an incremental path ending at the top-level with complex autonomous systems. In this layered organization, all behaviors have permanent access to the specific sensing facility and compete in gaining control over the effector. In order to achieve a reasonable global behavior, a ranking of importance is considered for all behaviors, and only the most important ones have a chance to become active. The relevant behavior or subset of behaviors are triggered on occasion of specific sensations in the environment. For example, the obstacle-avoiding behavior must become active before collision, in order to guarantee the survival of the robot, otherwise the original task-solving behavior would be active.

The characteristics of autonomous robot systems, i.e. situatedness, corporeality, competence, emergence, and behavioral organization, have been formulated at an abstract level. For the concrete development of autonomous *camera-equipped* robot systems, one must lay the appropriate foundations. We propose, that the characteristics of Robot Vision just make up this foundation. They have been summarized in Subsection 1.2.2 by animated attention, purposive perception, visual demonstration, compatible perception, biased learning, and feedback analysis. Figure 1.1 shows the characteristics of an autonomous robot system (top level) together with the characteristics of Robot Vision (bottom level), which together characterize an *autonomous camera-equipped robot system*. As learned in the discussion of alternative purposes of robots (Subsection 1.3.1), an autonomous camera-equipped robot system can be a robot-supported vision system, a vision-supported robot system, or a combination of both.

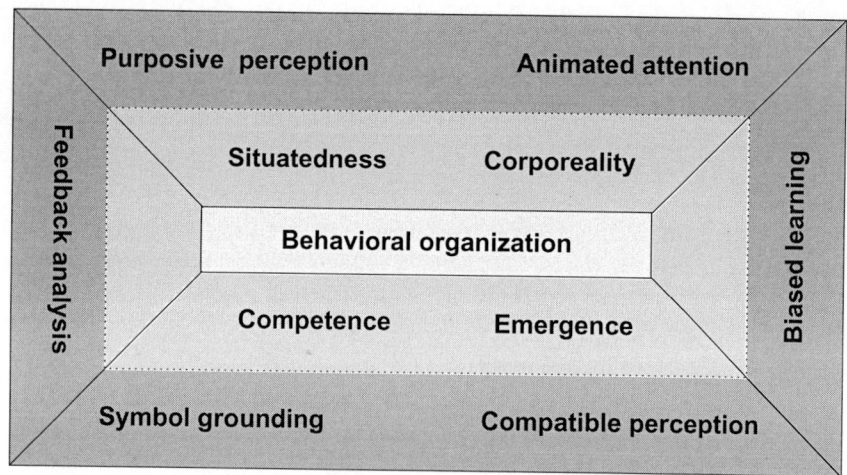

Fig. 1.1. Characterization of autonomous camera-equipped robot systems.

1.3.3 Autonomous Camera-Equipped Robot System

Definition 1.4 (Autonomous camera-equipped robot system) *An autonomous camera-equipped robot system is a robot system including robot heads and/or cameras, which shows autonomous task-solving behaviors of visual perception and/or action. The basis for autonomy is situatedness, corporeality, competence, and emergence. These characteristics can be reached by animated attention, purposive perception, visual demonstration, compatible perception, biased learning, and feedback analysis.*

All characteristics of an autonomous camera-equipped robot system are highly correlated to each other. Any one could be taken as seed from which to evolve the others. This becomes obvious in the next Section 1.4, in which the role of visual demonstration is explained with regard to learning in reality.

1.4 Important Role of Demonstration and Learning

In an autonomous camera-equipped robot system, the 3D spatial relation between effector, camera(s), and/or object(s) must change according to a task-relevant strategy. The images produced by camera(s) are input of autonomy-relevant functions which are responsible for generating appropriate control signals for the effectors.

It is obvious that a 3D world situation between objects or between object and effector appears differently in the 2D images in the case of varying viewing positions [96] or varying lighting conditions [17]. Conversely, different 3D world situations can lead to similar 2D images (due to loss of one dimension) or dissimilar 2D images. Therefore, classes of image feature values must be determined which originate from a certain 3D situation or a certain set of 3D situations. In this work, two types of classes will be distinguished, and, related to them, the concepts of *compatibilities* and *manifolds* will be introduced.

> Compatibilities describe constraints on the process of image formation which do hold, more or less, under task-relevant or random variations of the imaging conditions.

The class of image feature values involved in a compatibility is represented by a representative value together with typical, small deviations.

> Manifolds describe significant variations of image feature values which originate under the change of the 3D spatial relation between effector, camera, and/or objects.

The class of image feature values involved in a manifold is represented extensively by basis functions in a canonical system. Visual demonstration is the basis for learning compatibilities and manifolds in the real task-relevant world. Generally, the learned compatibilities can be used for parameterizing prior vision algorithms and manifolds can be used for developing new vision algorithms.

For the vision-based control of the effector, the relationship between environment, effector, and image coordinates must be determined. Specifically, the purpose is to transform image coordinates of objects into the coordinate system of the effector. The autonomy-relevant functions for effector control are based on the combined use of feature compatibilities, feature manifolds, and environment-effector-image relationships.

1.4.1 Learning Feature Compatibilities under Real Imaging

By eliciting fundamental principles underlying the process of image formation, one can make use of a generic bias, and thus reduce the role of object-specific knowledge for structure extraction and object recognition in the image [117, pp. 26-30]. Theoretical assumptions (*e.g.* projective invariants) concerning the characteristic of image formation which can be proven nicely for simulated pinhole cameras, generally do not hold in practical applications. Instead, realistic qualitative assumptions (so-called *compatibilities*) must be learned in an offline phase prior to online application.

Compatibility of Regularities under Geometric Projection

Shape descriptions are most important if they are *invariants* under geometric projection and change of view. First, the perceptual organization of line segments into complex two-dimensional constructs, which originate from the surface of three-dimensional objects, can be based on invariant shape regularities. For example, simple constructs of parallel line segment pairs are used by Ylä-Jääski and Ade [179], or sophisticated constructs of repeated structures or rotational symmetries by Zisserman *et al.* [182]. Second, invariant shape regularities are constant descriptions of certain shape classes and, therefore, can be used as indices for recognition [143]. A real camera, however, executes a projective transformation in which shape regularities are relaxed in the image, *e.g.* three-dimensional symmetries are transformed into two-dimensional skewed symmetries [77]. More generally, projective quasi-invariants must be considered instead of projective invariants [20].

By demonstrating sample objects including typical regularities and visually perceiving the objects using actual cameras, one can make measurements of real deviations from regularities (two-dimensionally projected), and thus learn the relevant degree of compatibility.

Compatibility of Object Surface and Photometric Invariants

Approaches for recognition and/or tracking of objects in images are confronted with variations of the gray values, caused by changing illumination conditions. The object illumination can change directly with daylight and/or the power of light bulbs, or can change indirectly by shadows arising in the spatial relation between effector, camera, and object. The problem is to convert color values or gray values, which depend on the illumination, into descriptions that do not depend on the illumination. However, solutions for perfect color constancy are not available in realistic applications [55], and therefore approximate photometric invariants are of interest. For example, normalizations of the gray value structure by standard or central moments of second order can improve the reliability of correlation techniques [148].

By demonstrating sample objects under typical changes of the illumination one can make measurements of real deviations from exact photometric invariants, and thus learn the relevant degree of compatibility.

Compatibility of Geometric and Photometric Image Features

The general assumption behind all approaches of object detection and boundary extraction is that three-dimensional surface discontinuities should have corresponding gray value edges in the image.[4] Based on this, a compatibility between the geometric and photometric type of object representation must hold in the image. For example, the orientation of an object boundary line in the image must be similar to the orientation of a gray value edge of a point on the line [135]. A further example, the junction angle of two boundary lines must be similar to the opening angle of the gray value corner at the intersection point of the lines. The geometric line features are computed globally in an extended patch of the image, and the photometric edge or corner feature are computed locally in a small environment of a point. Consequently, by the common consideration of geometric and photometric features one also verifies the compatibility between global and local image structure.

By demonstrating sample objects including typical edge curvatures and extracting geometric and photometric image features, one can compare the real measurements and learn the relevant degree of compatibility.

Compatibility of Motion Invariants and Changes in View Sequence

In an autonomous camera-equipped robot system, the spatial relation between camera(s) and object(s) changes continually. The task-solving process could be represented by a discrete series of changes in this spatial relation, *e.g.* one could consider the changing relations for the task of moving the robot hand of a manipulator towards a target object while avoiding obstacle objects. Usually, there are different possibilities of taking trajectories subject to the constraint of solving the task. A cost function must be used for determining the cheapest course. Beside the typical components of the cost function, *i.e.* distance to goals and obstacles, it must also include a measure of difficulty of extracting and tracking task-relevant image features. This aspect is directly related with the computational effort of image sequence analysis and, therefore, has influence on the real-time capability of an autonomous robot system. By constraining the possible camera movements appropriately, the flow vector fields originating from scene objects are easy to represent. For example, a straight camera movement parallel over a plane face of a three-dimensional object should reveal a uniform flow field at the face edges. A further example, if a camera is approaching an object or is rotating around the optical axis which is normal to the object surface, then *log-polar transformation (LPT)* can be applied to the gray value images. The motivation lies in the fact that during the camera movement, simple shifts of the transformed object pattern occur without any pattern distortions [23]. However, in the view sequence these invariants only hold for a simulated pinhole camera

---

[4] See the book of Klette et al. [90] for geometric and photometric aspects in Computer Vision.

whose optical axis must be kept accurate normal to the object surface while moving the camera.

By demonstrating sample objects and executing typical camera movements relative to the objects, one can make measurements of real deviations from uniformity of the flow field in original gray value or in transformed images, and thus learn the relevant degree of compatibility between 3D motions and 2D view sequences.

Invariants Are Special Cases of Compatibilities

In classical approaches of Computer Vision, invariants are constructed for a group of transformations, *e.g.* by eliminating the transformation parameters [111]. In real applications, however, the actual transformation formula is not known, and for solving a certain robot task only a relevant subset of transformations should be considered (possibly lacking characteristics of a group). The purpose of visual demonstration is to consider the real corporeality of robot and camera by learning realistic compatibilities (involved in the imaging process) instead of assuming non-realistic invariants. Mathematically, a compatibility must be attributed with a statistical probability distribution, which represents the probabilities that certain degrees of deviation from a theoretical invariant might occur in reality. In this work, Gaussian probability distributions are considered, and based on that, the Gaussian extent value σ can be used to define a confidence value for the adequacy of a theoretical invariant. The lower the value of σ, the more confident is the theoretical invariant, *i.e.* the special case of a compatibility with σ equal to 0 characterizes a theoretical invariant. In an experimentation phase, the σ values of interesting compatibilities are determined by visual demonstration and learning, and in the successive application phase, the learned compatibilities are considered in various autonomy-relevant functions. This methodology of acquiring and using compatibilities replaces the classical concept of non-realistic, theoretical invariants.

The first attempt of relaxing invariants has been undertaken by Binford and Levitt, who introduced the concept of quasi-invariance under transformations of geometric features [20]. The compatibility concept in our work is a more general one, because more general transformations can be considered, maybe with different types of features prior and after the mapping.

1.4.2 Learning Feature Manifolds of Real World Situations

For the detection of situations in an image, *i.e.* in answer to the question *"Where is which situation ?"* [12], one must acquire models of target situations in advance. There are two alternatives for acquiring such model descriptions. In the first approach, detailed models of 3D target situations and projection functions of the cameras are requested from the user, and from that the relevant models of 2D target situations are computed [83]. In many real world applications, however, the gap between 2D image and 3D world situations is problematic, *i.e.* it is difficult, costly, and perhaps even impossible to obtain realistic 3D models and realistic projection functions.[5] In the second approach, descriptions of 2D target situations are acquired directly from image features based on visual demonstration of 3D target situations and learning of feature manifolds under varying conditions [112]. For many tasks to be carried out in typical scenes, this second approach is preferable, because actual objects and actual characteristics of the cameras are considered directly to model the 2D target situations. A detection function must localize meaningful patterns in the image and classify or evaluate the features as certain model situations. The number of task-relevant image patterns is small in proportion to the overwhelming number of all possible patterns [136], and therefore a detection function must represent the manifolds of relevant image features implicitly.

In the following we use the term *feature* in a general sense. An image pattern or a collection of elementary features extracted from a pattern will simply be called a feature. What we really mean by a feature is a vector or even a complicated structure of elementary (scalar) features. This simplification enables easy reading, but where necessary we present the concrete specification.

Learning Feature Manifolds of Classified Situations

The classification of a feature means assigning it to those model situation whose *feature manifold* contains the relevant feature most appropriately, *e.g.* recognize a feature in the image as a certain object. Two criteria should be considered simultaneously, robustness and efficiency, and a measure is needed for both criteria in order to judge different feature classifiers. For the robustness criterion, a measure can be adopted from the literature on statistical learning theory [170, 56] by considering the definition of *probably approximately correct learning, (PAC-learning)*. A set of model situations is said to be PAC-learned if, with a probability of at least P^r, a maximum percentage E of features is classified erroneous. Robustness can be defined reasonably by the quotient of P^r by E, *i.e.* the higher this quotient, the more robust is the classifier. It is conceivable that high robustness requires an extensive amount of attributes for describing the classifier. In order, however, to reduce the computation effort of classifying features, a *minimum description length* of the classifier is prefered [139]. For the obvious conflict between robustness and efficiency a compromise is needed.

---

[5] Recent approaches of this kind use more general, parametric models which express certain unknown variabilities, and these are verified and fine-tuned under the actual situations in the images [97]. With regard to the qualitativeness of the models, these new approaches are similar to the concept of compatibilities in our work, as discussed above.

By demonstrating appearance patterns of *classified situations*, one can experimentally learn several versions of classifiers and finally select the one which carries out the best compromise between robustness and efficiency.

Learning Feature Manifolds of Scored Situations

Task-relevant changes of 3D spatial relations between effector, camera(s), and/or object(s) must be controlled by assessing for the stream of images the successive 2D situations relative to the 2D goal situation. The intermediate situations are considered as discrete steps in a course of *scored situations* up to the main goal situation. Classified situations (see above) are a special case of scored situations with just two possible scores, *e.g.* values 0 or 1. In the continual process of robot servoing, *e.g.* for arranging, grasping, or viewing objects, the differences between successive 2D situations in the images must correlate with certain changes between successive 3D spatial relations. Geometry-related features in the images include histograms of edge orientation [148], results of line Hough transformation [123], responses of situation-specific Gabor filters [127], etc. Feature manifolds must characterize scored situations, *e.g.* the gripper is 30 percent off from the optimal grasping situation. Both for learning and applying these feature manifolds the coherence of situations can be taken into account. A course of scored situations is said to be PAC-learned if, with a probability of at least P^r, a maximum deviation D from the actual score is given.

By demonstrating appearance patterns of scored situations, one can learn several versions of scoring modules and finally select the best one. Experiments with this scoring module are relevant for determining the degree of correlation between scored situations in the images and certain 3D spatial relations.

Systems of Computer Vision include the facility of geometric modeling, *e.g.* by operating with a CAD subsystem (computer aided design). The purpose is to incorporate geometric models, which are needed for the recognition of situations in the images. However, in many realistic applications one is asking too much of the system user if requested to construct the models for certain target situations. By off-line visual demonstration of situations under varying, task-relevant conditions, the model situations can be acquired and represented as manifolds of image features directly. In this work, an approach for learning manifolds is presented which combines Gaussian basis function networks [21, pp. 164-193] and principal component analysis [21, pp. 310-319].

The novelty of the approach is that the coherence of certain situations will be used for constraining (local and global) the complexity of the appearance manifold. The robustness improves both in recognition and scoring functions.

1.4.3 Learning Environment-Effector-Image Relationships

The effector interacts with a small environmental part of the world. For manipulating or inspecting objects in this environment, their coordinates must be determined relative to the effector. The relationship between coordinates in the image coordinate system and the effector coordinate system must be formalized [173, 78]. The relevant function can be learned automatically by controlled effector movement and observation of a calibration object.

Learning Relationships for an Eye-off-Hand System

For an eye-off-hand system, the gripper of a manipulator can be used as calibration object which is observed by cameras without physical connection to the robot arm. The gripper is steered by a robot program through the working space, and the changing image and manipulator coordinates of it are used as samples for learning the relevant function.

Learning Relationships for an Eye-on-Hand System

For an eye-on-hand system the camera(s) is (are) fastened on the actuator system for controlling inspection or manipulation tasks. A natural or artificial object in the environment of the actuator system serves as calibration object. First, the effector is steered by the operator (manually using the control panel) into a certain relation to this calibration object, *e.g.* touching it or keeping a certain distance to it. In this way, the goal relation between effector and an object is stipulated, something which must be known in the application phase of the task-solving process. Specifically, a certain environmental point will be represented more or less accurately in actuator coordinates. Second, the effector is steered by a robot program through the working space, and the changing image coordinates of the calibration object and position coordinates of the effector are used as samples for learning the relevant function.

These strategies of learning environment-effector-image relationships are advantageous in several aspects. First, by controlled effector movement, the relevant function of coordinate transformation is learned directly, without computing the intrinsic camera parameters and avoiding artifical coordinate systems (*e.g.* external world coordinate system). Second, the density of training samples can easily be changed by different discretizations of effector movements. Third, a natural object can be used instead of an artificial calibration pattern. Fourth, task-relevant goal relations are demonstrated instead of modeling them artificially.

The learned function is used for transforming image coordinates of objects into the coordinate system of the actuator system.

1.4.4 Compatibilities, Manifolds, and Relationships

A 3D spatial relation between effector and objects appears differently in the image depending on viewing and lighting conditions. Two concepts have been mentioned for describing the ensemble of appearance patterns, *i.e.* compatibilities and manifolds.

Extracting Reliable and Discriminative Image Features

By uncovering compatibilities under real image formation, one determines
parameters and constraints for prior vision algorithms. This method partially regularizes the ill-posed problem of feature extraction [130]. A principal
purpose of learning and applying compatibilities is to obtain more reliable
features from the images without taking specific object models into account.
Furthermore, certain compatibilities can lead to the extraction of image features which are nearly constant for the ensemble of appearance patterns of
the 3D spatial relation, *i.e.* the features are a common characterization of
various appearances of the 3D spatial relation. However, the extracted image features may not be discriminative versus appearance patterns which
originate from other 3D spatial relations.

Specifically, this aspect plays a significant role in the concept of manifolds. Appearance patterns from different 3D spatial relations must be distinguished, and the various patterns of an individual 3D relation should be
collected. Manifolds of image features are the basis for a robust and efficient
discrimination between different 3D situations, *i.e.* by constructing recognition or scoring functions. The features considered for the manifolds have an
influence on the robustness, and the complexity of the manifolds has an influence on the efficiency of the recognition or scoring functions. Therefore,
the results from learning certain compatibilities are advantageous for shaping
feature manifolds. For example, by applying gray value normalization in a relevant window (compatibility of object surface and photometric invariants),
one reduces the influence of lighting variations in the appearance patterns,
and, consequently, the complexity of the pattern manifold is simplified.

Role of Active Camera Control

Both the complexity of manifolds and the validity of compatibilities are affected by constraints in spatial relations between effector, camera(s), and/or
object(s). A first category of constraints is fixed by the corporeality of the
camera-equipped robot system, *i.e.* the effectors, mechanics, kinematics, objectives, etc. A second category of constraints is flexible according to the
needs of solving a certain task. Two examples of constraints are mentioned
regarding the relationship of the camera in the environment. First, camera
alignment is useful for reducing the complexity of manifolds. By putting the
camera of an eye-off-hand system in a standard position and/or orientation,
one reduces the variety of different camera-object relations. In this case, the
complexity of the appearance manifold depends only on position and/or illumination variations of the objects in the working environment. Second,
controlled camera motion is useful in fulfilling compatibilities. For a detailed
object inspection, the camera should approach closely. If the optical axis is
kept normal to the object surface while moving the camera and applying
log-polar transformation to the gray value images, then the resulting object pattern is merely shifting but not increasing. These two examples make

clear that the concepts of manifolds and compatibilities are affected by the animated vision principle. Cameras should be aligned appropriately for reducing the complexity of manifolds, and desired compatibilities can be taken as constraints for controlling the camera movement.

Compatibilities and Manifolds in Robot Vision

The two concepts of compatibilities and manifolds are essential to clarify the distinction between Computer Vision and Robot Vision. The concept of invariance is replaced by the more general concept of compatibility, and the concept of geometric models is subordinated to feature manifolds. Both the compatibilities and the manifolds must be learned by (visual) demonstrations in the task-relevant environment. Actually, these principles contribute substantially to the main issues arising in Computer Vision (see Section 1.2), *i.e.* origin of models and strategy of model application. In recent years, several workshops have been organized dedicated to performance characteristics and quality of vision algorithms.[6] In the paradigm of Robot Vision, the task-relevant environment is the origin respective justification of the vision algorithms, and therefore it is possible to assess their actual quality.

1.5 Chapter Overview of the Work

Chapter 2 presents a novel approach for localizing a three-dimensional target object in the image and extracting the two-dimensional polyhedral depiction of the boundary. Polyhedral descriptions of objects are needed in many tasks of robotic object manipulation, *e.g.* grasping and assembling tasks. By eliciting the general principles underlying the process of image formation, we extensively make use of general, qualitative assumptions, and thus reduce the role of object-specific knowledge for boundary extraction. *Geometric/photometric compatibility* principles are involved in an approach for extracting line segments which is based on Hough transformation. The perceptual organization of line segments into polygons or arrangements of polygons, which originate from the silhouette or the shape of approximate polyhedral objects, is based on shape regularities and compatibilities of projective transformation. An affiliated saliency measure combines evaluations of geometric/photometric compatible features with geometric grouping features. An ordered set of most salient polygons or arrangements is the basis for locally applying techniques of object recognition (see Chapter 3) or detailed boundary extraction. The generic approach is demonstrated for technical objects of electrical scrap located in real-world cluttered scenes.

Chapter 3 presents a novel approach for object or situation recognition, which does not require a priori knowledge of three-dimensional geometric

---

[6] Interesting papers also have been collected in a special issue of the journal *Machine Vision and Applications* [37].

shapes. Instead, the task-relevant knowledge about objects or situations is grounded in photometric appearance directly. For the task of recognition, the appropriate *appearance manifolds* must be learned on the basis of visual demonstration. The feature variation in a manifold is specified algebraically by an implicit function for which certain deviations from ideal value 0 are accepted. Functions like these will serve as operators for the recognition of objects under varying view angle, view distance, illumination, or background, and also serve as operators for the recognition of scored situations. Mixtures of Gaussian basis function networks are used in combination with Karhunen-Loéve expansion for function approximation. Several architectures will be considered under the trade-off between efficiency, invariance, and discriminability of the recognition function. The versions take care for correlations between close consecutive view patterns (local in the manifold) and/or for the relationship between far distant view patterns (global in the manifold). The greatest strength of our approach to object recognition is the ability to learn compatibilities between various views under real world changes. Additionally, the compatibilities will have discriminative power versus counter objects.

Chapter 4 uses the dynamical systems theory as a common framework for the design and application of camera-equipped robots. The matter of dynamics is the changing relation between robot effectors and environmental objects. We present a *modularized dynamical system* which enables a seamless transition between designing and application phase and an uniform integration of reactive and deliberate robot competences. The designing phase makes use of bottom-up methodologies which need systematic experiments in the environment and apply learning and planning mechanisms. Task-relevant deliberate strategies and parameterized control procedures are obtained and used in the successive application phase. This online phase applies visual feedback mechanisms which are the foundation of vision-based robot competences. A layered configuration of dynamic vector fields uniformly represents the task-relevant deliberate strategies, determined in the designing phase, and the perception-action cycles, occuring in the application phase. From the algorithmic point of view the outcome of the designing phase is a configuration of concrete modules which is expected to solve the underlying task. For this purpose, we present three categories of generic modules, i.e. instructional, behavioral, and monitoring modules, which serve as design abstractions and must be implemented for the specific task. The resulting behavior in the application phase should meet requirements like task-relevance, robustness, flexibility, time limitation, *etc.* simultaneously. Based on a multi-component robotic equipment the system design and application is shown exemplary for a high-level task, which includes sub-tasks of localizing, approaching, grasping, and carrying objects.

Chapter 5 summarizes the work and discusses future aspects of behavior-based robotics.

2. Compatibilities for Object Boundary Detection

This chapter presents a generic approach for object localization and boundary extraction which is based on the extensive use of feature compatibilities.

2.1 Introduction to the Chapter

The introductory section of this chapter embeds our methodology of object localization and boundary extraction in the general context of purposive, qualitative vision, then presents a detailed review of relevant literature, and finally gives an outline of the following sections.[1]

2.1.1 General Context of the Chapter

William of Occam (ca. 1285-1349) was somewhat of a minimalist in medieval philosophy. His motto, known as *Occam's Razor*, reads as follows: *"It's vain to do with more what can be done with less."* This economy principle (opportunism principle) is self-evident in the paradigm of *Purposive Vision* [2].

---

1. The vision system should gather from the images only the information relevant for solving a specific task.
2. The vision procedures should be generally applicable to a category of similar tasks instead of a single specific task.

Ad 1. In the field of vision-supported robotics, an example for a category of similar tasks is the robot grasping of technical objects. In most cases a polyhedral approximation of the target object is sufficient (see the survey of robot grasp synthesis algorithms in [156]). For example, in order to grasp an object using a parallel jaw gripper, it is sufficient to reconstruct from the image a rectangular solid, although the object may have round corners or local protrusions. The corporeal form of the robot gripper affects the relevant type of shape approximation of the object, *i.e.* purposive, qualitative description of the geometric relation between gripper und object.

---

[1] The extraction of solely purposive information is a fundamental design principle of Robot Vision, see Chapter 1.

J. Pauli: Learning-Based Robot Vision, LNCS 2048, pp. 25-99, 2001.
© Springer-Verlag Berlin Heidelberg 2001

For the qualitative reconstruction of these shapes, a certain limited spectrum of image analysis tools is useful which additionally depends on the characteristic of the camera system. For example, if a lens with large focal length is applied, then it is plausible to approximate the geometric aspect of image formation by perspective collineations [53]. As straight 3D object boundary lines are projected into approximate straight image lines, one can use techniques for straight line extraction, *e.g.* Hough transformation [100]. Altogether, the kind of task determines the *degree of qualitativeness* of the information that must be recovered from the images.[2] A restriction to *partial recovery* is inevitable for solving a robot task in limited time with minimum effort [9]. In this sense, the designing of autonomous robot systems drives the designing of included vision procedures (see Chapters 1 and 4, and the work of Sommer [161]).

Ad 2. Applying Occam's Razor to the designing of vision procedures also means a search for applicability for a *category of similar tasks* instead of a single specific task. This should be reached by exploiting *ground truths* concerning the situatedness and the corporeality of the camera-equipped robot system. The ground truths are constraints on space-time, the camera system, and their relationship, which can be generally assumed for the relevant category of robot tasks. General assumptions of various types have been applied more or less successfully in many areas of image processing and Computer Vision.

a) Profiles of step, ramp, or sigmoid functions are used as *mathematical models* in procedures for edge detection [72].

b) For the perceptual organization of edges into structures of higher complexity (*e.g.* line segments, curve segments, ellipses, polygons), approaches of edge linking are applied which rely on *Gestalt principles* of proximity, similarity, closure, and continuation [103].

c) Object recognition is most often based on geometric quantities which are assumed to be *invariant* under the projective transformations used to model the process of image formation [110]. Usually, a stratification into euclidean, similarity, affine, and projective transformations is considered with the property that in this succession each group of transformations is contained in the next. The sets of invariants of these transformation groups are also organized by subset inclusion but in reverse order, *e.g. cross-ratio* is an invariant both under projective and euclidean transformation, whereby length is only invariant under euclidean transformations.

d) Frequently, in real applications the assumption of invariance is too strong and must be replaced by the assumption of *quasi-invariance*. Its theory and important role for grouping and recognition has been worked out by Binford and Levitt [20]. For example, Gros *et al.* use geometric quasi-invariants of pairs of line segments to match and model images [68].

---

[2] See also the *IEEE Workshop on Qualitative Vision* for interesting contributions [4].

e) Finally, for the reconstruction of 3D shape and/or motion the ill-posed problem is treated using regularization approaches, which incorporate *smoothness and rigidity assumptions* of the object surface [52].

Problems with the Use of Knowledge in Image Analysis

A critical introspection reveals some problems concerning the applicability of all these constraints. For example, Jain and Binford have pointed out that smoothness and rigidity constraints of objects must be applied locally to image regions of depicted object surfaces, but the major problem is to find those meaningful areas [86]. Obviously, in real applications the listed assumptions are too general, and should be more directly related to the type of the actual vision task, *i.e.* the categories of relevant situations and goals. In knowledge-based systems for image understanding (*e.g.* the system of Liedtke *et al.* [102]), an extensive use of domain-specific knowledge was proposed [43]. These systems fit quite well to Marr's theory of vision in the sense of striving for *general vision systems* by explicitly incorporating object-specific assumptions [105]. However, the extensive use of knowledge contradicts Occam's economy principle, and in many applications the explicit formulation of object models is difficult and perhaps even impossible.

2.1.2 Object Localization and Boundary Extraction

Having the purpose of robotic grasping and arranging in mind, we present a system for localizing approximate polyhedral objects in the image and extracting their qualitative boundary line configurations. The approach is successful in real-world robotic scenes which are characterized by clutter, occlusion, shading, *etc.* A *global-to-local strategy* is favoured, *i.e.* first to look for a candidate set of objects by taking only the approximate silhouette into account, then to recognize target objects of certain shape classes in the candidate set by applying view-based approaches, and finally to extract a detailed boundary.

Extracting Salient Polygons or Arrangements of Polygons

Our approach for localization is to find *salient polygons*, which represent single faces or silhouettes of objects. The saliency of polygons is based on *geometric/photometric compatible features* and on *geometric regularity features*. The first category of features comprises compatibility evaluations between geometric and photometric *line features* and geometric and photometric *junction features*. The Hough transformation is our basic technique for extracting line segments and for organizing them into polygons. Its robustness concerning parameter estimation is appreciated and the loss of locality is overcome by the geometric/photometric compatibility principles. For extracting and characterizing junctions, a corner detector is used in combination with a rotating wedge filter. The second category of features involved in the saliency measure of polygons comprises *geometric compatibilities* under projective transformation. Specifically, they describe regularity aspects of 3D silhouettes and 2D polygons, respectively. Examples for regularities are *parallelism, right-angles, reflection-symmetry, translation-symmetry.*

The ordered places of most salient polygons are visited for special local treatment. First, an appearance-based approach can be applied for specific object recognition or recognition of certain shape classes (see Chapter 3).[3] Second, a generic procedure can be applied for detailed boundary extraction of certain shape classes, *e.g.* parallelepipeds. Our approach is to extract arrangements of polygons from the images by incorporating a *parallelism compatibility*, a *pencil compatibility*, and a *vanishing-point compatibility*, all of which originate from general assumptions of projective transformation of regular 3D shapes. A major contribution of this chapter is that the basic procedure of line extraction, *i.e.* Hough transformation, and all subsequent procedures are controlled by constraints which are inherent in the *three-dimensional nature* of the scene objects and inherent in the *image formation principles* of the camera system.

General Principles instead of Specific Knowledge

Our system is organized in several procedures for which the relevant assumptions are clearly stated. The assumptions are related to the situatedness and corporeality of the camera-equipped robot system, *i.e.* compatibilities of regular shapes under projective transformation and geometric/photometric compatibilities of image formation. Furthermore, these assumptions are stratified according to *decreasing generality*, which imposes a certain degree of generality on the procedures. Concerning the objects in the scene, our most general assumption is that the object shape is an approximate polyhedron, and an example for a specific assumption is that an approximate parallelepiped is located in a certain area. We follow the claims of Occam's minimalistic philosophy and elicit the general principles underlying the perspective projection of polyhedra, and then implement procedures as generally applicable as possible. Based on this characterization of the methodology, relevant contributions in the literature will be reviewed.

2.1.3 Detailed Review of Relevant Literature

Object detection can be considered as a cyclic two-step procedure of localization and recognition [12], which is usually organized in several levels of data abstraction. Localization is the task of looking for image positions where objects of a certain class are located. In the recognition step, one of these locations is considered to identify the specific object. Related to the problem

---

[3] Alternatively, in certain applications histogram-based indexing approaches are also useful [164].

of boundary extraction, the task of localization is strongly correlated to perceptual organization, *e.g.* to organize those gray value edges which belong to the boundary of a certain object.

Approaches to Perceptual Organization

Sarkar and Boyer have reviewed the relevant work (up to year 1992) in *perceptual organization* [146] and proposed a four-level classification of the approaches, *i.e. signal level, primitive level, structural level, assembly level*. For example, at the signal level pixels are organized into edge chains, at the primitive level the edge chains are approximated as polylines (*i.e.* sequences of line segments), at the structural level the polylines are combined to polygons, and at the assembly level several polygons are organized into arrangements. For future research they suggested:

> *"There is a need for research into frameworks for integration of various Gestaltic cues including non-geometric ones ..."*

Sarkar and Boyer also presented a hierarchical system for the extraction of curvilinear or rectilinear structures [147]. Regularities in the distribution of edges are detected using "voting" methods for *Gestaltic phenomena* of proximity, similarity, smooth continuity and closure. The approach is generic in the sense that various forms of tokens can be treated and represented as graphs, and various types of structures can be extracted by applying standardized graph analysis algorithms. Our approach incorporates several types of non-geometric cues (*i.e.* photometric features), treats closed line configurations of higher complexity including higher level Gestaltic phenomena, and from that defines a saliency measure for different candidates of line organizations.

In a work of Zisserman *et al.* grouping is done at all four levels [182]. Line structures belonging to an object are extracted by using techniques of edge detection, contour following and polygonal approximation (signal level, primitive level, structural level). The representation is given by certain invariants to overcome difficulties in recognizing objects under varying viewpoints. These geometric invariants are used to define an indexing function for selecting certain models of object shapes, *e.g.* certain types of polyhedra or surfaces of revolution. Based on a minimal set of invariant features, a certain object model is deduced, and based on that, a class-based grouping procedure is applied for detailed boundary extraction (assembly level). For example, under affine imaging conditions the parallelism of 3D lines of a polyhedra also holds between the projected lines in the image. Accordingly, certain lines of the outer border of an object appear with the same orientation in the interior of the silhouette. This gives an evidence of grouping lines for describing a polyhedron. Our approach contributes to this work, in that we introduce some assembly level grouping criteria for boundary extraction of approximate polyhedra. These criteria are a parallelism compatibility, a pencil compatibility, and a vanishing-point compatibility.

Castano and Hutchinson present a probabilistic approach to perceptual grouping at the primitive and structural level [33]. A probability distribution over a space of possible image feature groupings is determined, and the most likely groupings are selected for further treatment. The probabilities are based on how well a set of image features fits to a particular geometric structure and on the expected noise in image data. The approach is demonstrated for two types of low-level geometric structures, *i.e.* straight lines and bilateral symmetries. Complex symmetrical structures consisting of groups of line segments are extracted in a work of Ylä-Jääski and Ade [179]. In a two-step procedure, pairs of line segments are first detected which are the basic symmetry primitives, and then several of them are selectively grouped along the symmetry axes of segment pairs. Our approach is more general in the sense that we treat further types of regularities and symmetries.

Amir and Lindenbaum present a grouping methodology for both signal and primitive levels [3]. A graph is constructed whose nodes represent primitive tokens such as edges, and whose arcs represent grouping evidence based on collinearity or general smoothness criteria. Grouping is done by finding the best graph partition using a maximum likelihood approach. A measure for the quality of detected edge organizations is defined which could be used as decision function for selectively postprocessing certain groups. In our approach, graph analysis is avoided at the primitive level of grouping edges (because it seems to be time-consuming), but is used for detecting polygons at the structural level.

Cho and Meer propose an approach for detecting image regions by evaluating a compatibility among a set of slightly different segmentations [36]. Local homogeneity is based on co-occurrence probabilities derived from the ensemble of initial segmentations, *i.e.* probabilities that two neighboring pixels belong to the same image region. Region adjacency graphs at several levels are constructed and exploited for this purpose. In our work, image segmentation is based on compatibilities between geometric and photometric features and on geometric regularity features which are compatible under projective transformation.

Hough Transformation as Possible Foundation

Frequently, *Hough transformation* has been used as basic procedure for grouping at the primitive or structural level [100]. Specific arrangements of gray value edges are voting for certain analytic shapes, *e.g.* straight lines or ellipses. For example, each line of edges creates a peak of votes in the space of line parameters, and the task is to localize the peaks. In order to make the Hough transformation more sensitive, one can go back to the signal level and take the orientations of the gray value edges into account. Since orientation is a parameter in a polar representation of a line, the number of possible line orientations, for which a pixel may vote, can be reduced to the relevant one. The size of this voting kernel influences the sharpness of the Hough peaks [123], *i.e.* the accuracy of line parameters. The Hough image can be used for grouping at the structural level or even at the assembly level. The problem is to find especially those peaks which arise from lines belonging to a specific object. Princen *et al.* use a hierarchical procedure which extracts an exhaustive set of peaks and afterwards selects the relevant subset by applying Gestaltic grouping criteria [135]. Wahl and Biland extract objects from a polyhedral scene by representing an object boundary as a distributed pattern of peaks in the parameter space of lines [172]. Alternatively, Ballard introduced the generalized Hough transformation for the extraction of complex natural 2D shapes, in which a shape is represented in tabular form instead of an analytic formula [10].

This short review of extensions of the standard Hough transformation gives the impression that our Hough voting procedure can serve as basis for perceptual organization at all perception levels and for integrationg cues from all levels. The greatest weakness of the standard Hough transformation is the loss of locality, *e.g.* a line can gain support from pixels anywhere along its length from image border to border. Therefore, two or more line segments may be misinterpreted as one line, or short line segments may be overlooked. Consequently, Yang *et al.* introduce a weighted Hough transformation, in which the connectivity of a line is measured in order to detect also short line segments [178]. Similarly, Foresti *et al.* extend the Hough transformation to labeled edges [57]. Each label corresponds to a line segment which is extracted by a classical line following procedure taking connectivity and straightness in the course of edges into account. Our approach to overcome this problem is to apply three principles which are related to the geometric/photometric compatibility. The first one is the line/edge orientation compatibility (mentioned above), the second one takes the position and characterization of gray value corners into account, and the third principle consists of checking the similarity of local phase features along the relevant line segment. Furthermore, locality of boundary extraction is reached by applying a *windowed Hough transformation* within the areas of most salient polygons.

2.1.4 Outline of the Sections in the Chapter

Section 2.2 recalls the definitions of standard Hough transformation and orientation-selective Hough transformation for line extraction. Geometric/ photometric compatible features are introduced, based on the principle of orientation compatibility between lines and edges, and on the principle of junction compatibility between pencils and corners.

Section 2.3 defines regularity features of polygons, *i.e.* parallelism or right-angle between line segments and reflection-symmetry or translation-symmetry between polylines. For these features, certain compatibilities exist under projective transformation. Furthermore, the principle of phase compatibility between parallel line segments (short *parallel*) on the one hand, and gray value ramps on the other hand is introduced. The regularity features are combined with the geometric/photometric compatible features in a generic procedure for extracting salient quadrangles or polygons.

Section 2.4 introduces grouping criteria at the assembly level, *i.e.* the vanishing-point compatibility and the pencil compatibility. These assembly level criteria are integrated with the compatibilities which hold at the signal, primitive and structural levels or between them. Two generic procedures are presented for extracting the arrangements of polygons for approximate polyhedra.

In Section 2.5, task-relevant visual demonstrations are taken into account for learning the degrees of the involved compatibilities, *e.g.* justification of the compatibilities for typical objects from scenes of electrical scrap.

Section 2.6 discusses the approach on the basis of all introduced compatibilities which are assumed to be inherent in the three-dimensional nature of the scene objects, and/or inherent in the image formation principles of the camera system.

2.2 Geometric/Photometric Compatibility Principles

Obviously, the general assumption behind all approaches of boundary extraction is that three-dimensional surface discontinuities must have corresponding gray value edges in the image. Nearly all problems can be traced back to a gap between the geometric and the photometric type of scene representation. This section introduces two examples of reasonable *compatibilities between geometric and photometric features, i.e.* orientation compatibility between lines and edges, and junction compatibility between pencils and corners. The geometric features are single straight lines and pencils of straight lines, respectively. Hough transformation is used as basic procedure for line extraction. The photometric features are gray value edges and corners. Gradient magnitudes are binarized for the detection of gray value edges and Gabor wavelet operators are applied for estimating the orientations of the edges. The SUSAN operator is used for the detection and a rotating wedge filter for the characterization of gray value corners [159, 158].

2.2.1 Hough Transformation for Line Extraction

For representing straight image lines, we prefer the polar form (see Figure 2.1) which avoids singularities. Let $\mathcal{P}$ be the set of discrete coordinate tuples $p := (x_1, x_2)^T$ for the image pixels of a gray value image $\mathcal{I}^G$ with I_w columns and I_h rows. A threshold parameter δ_1 specifies the permissible deviation from linearity for a sequence of image pixels.

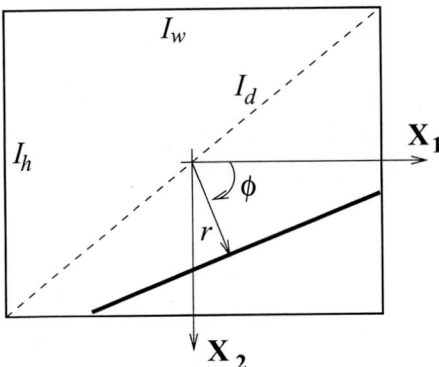

Fig. 2.1. Cartesian coordinate system with axes X_1 and X_2; Polar form of a line with distance parameter r and angle parameter ϕ taken relative to the image center.

Definition 2.1 (Polar representation of a line) *The polar representation of an image line is defined by*

$$f^L(p,q) := x_1 \cdot cos(\phi) + x_2 \cdot sin(\phi) - r, \quad |f^L(p,q)| \leq \delta_1 \qquad (2.1)$$

Parameter r is the distance from the image center to the line along a direction normal to the line. Parameter ϕ is the angle of this normal direction related to the x_1-axis. The two line parameters $q := (r, \phi)^T$ are assumed to be discretized. This calls for the inequality symbol in equation (2.1) to describe the permissible deviation from the ideal value zero. For the parameter space Q we define a discrete two-dimensional coordinate system (see Figure 2.2). The horizontal axis is for parameter r whose values reach from $\frac{-I_d}{2}$ to $\frac{+I_d}{2}$, whereby I_d is the length of the image diagonal. The vertical axis is for parameter ϕ whose values reach from $0°$ to $180°$ angle degrees. The discrete coordinate system can be regarded as a matrix consisting of I_d columns and 180 rows.

Due to discretization, each parameter tuple is regarded as a bin of real-valued parameter combinations, *i.e.* it represents a set of image lines with similar orientation and position. The *standard Hough transformation* counts for each bin how many edges in a gray value image lie along the lines which are specified by the bin. In the *Hough image* these numbers of edges are represented for each parameter tuple. Each *peak* in the Hough image indicates that the gray value image $\mathcal{I}^G$ contains an approximate straight line of edges, whose parameters are specified by the position of the peak. A binary image $\mathcal{I}^B$ is used which represents the edge points of $\mathcal{I}^G$ by $\mathcal{I}^B(p) = 1$ and all the other points by $\mathcal{I}^B(p) = 0$.

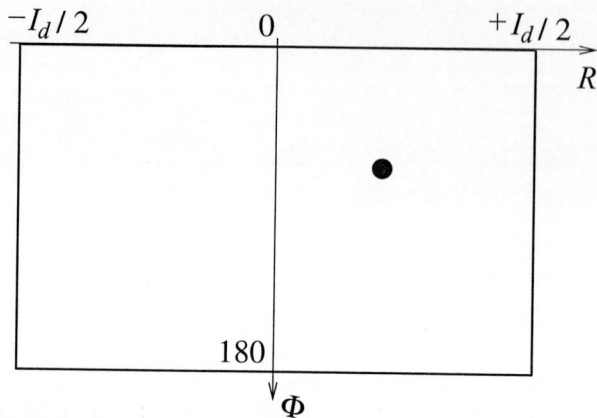

Fig. 2.2. Coordinate system for the two-dimensional space of line parameters (Hough image); Horizontal axis R for parameter r reaching from $-\frac{I_d}{2}$ to $+\frac{I_d}{2}$, and vertical axis Φ for parameter ϕ reaching from 0 to 180.

Definition 2.2 (Standard Hough transformation, SHT) *The standard Hough transformation (SHT) of the binary image $\mathcal{I}^B$ relative to the polar form f^L of a straight line is of functionality $\mathcal{Q} \rightarrow [0, \cdots, (I_w \cdot I_h)]$. The resulting Hough image $\mathcal{I}^{SH}$ is defined by*

$$\mathcal{I}^{SH}(q) := \# \left\{ p \in \mathcal{P} \mid \mathcal{I}^B(p) = 1 \ \wedge \ |f^L(p,q)| \leq \delta_1 \right\} \tag{2.2}$$

with symbol $\#$ denoting the number of elements of a set. Figure 2.3 shows on top the gray value image of a dark box (used as dummy box in electrical equipment) and at bottom the binarized image of gray value edges. The Hough image $\mathcal{I}^{SH}$ of the standard Hough transformation is depicted in Figure 2.4. Typically, wide-spread maxima occur due to the reason that all edges near or on a line cause the SHT to not only increase the level of the relevant bin but also many in their neighborhood. We are interested in sharp peaks in order to easy locate them in the Hough image (*i.e.* extract the relevant lines from the gray value image) and estimate the line parameters accurately. By making the discretization of the parameter space more fine-grained, the maxima would be more sharpened and more accurate. However, the computational expenditure for the Hough transformation would increase significantly.

2.2.2 Orientation Compatibility between Lines and Edges

The conflict between accuracy of extracted lines and efficiency of line extraction can be reduced by making use of an orientation compatibility between lines and gray value edges.

Fig. 2.3. (Top) Gray value image of an electrical dummy box; (Bottom) Binarized gradient magnitudes indicating the positions of gray value edges.

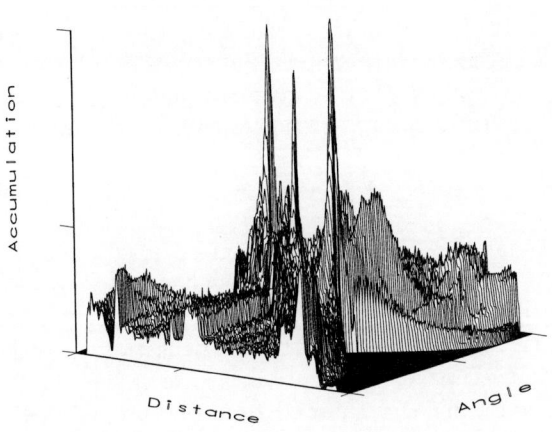

Fig. 2.4. Standard Hough transformation of the binarized image in Figure 2.3, *i.e.* accumulation array for discrete line parameters (distance r and angle ϕ). Widespread maxima in the Hough image (except for three sharp peaks).

Assumption 2.1 (Line/edge orientation compatibility for a line point)
The orientation ϕ of a line of gray value edge points and orientation $\mathcal{I}^O(p)$ of an edge at point $p := (x_1, x_2)^T$ on the line are approximately equal. The replacement of ϕ by $\mathcal{I}^O(p)$ in the polar form of the image line

implies just a small deviation from the ideal value zero. The necessary geometric/photometric compatibility is specified by parameter $\check{\delta}_2$.

$$|x_1 \cdot \cos(\mathcal{I}^O(x_1, x_2)) + x_2 \cdot \sin(\mathcal{I}^O(x_1, x_2)) - r| \leq \check{\delta}_2 \qquad (2.3)$$

A line of edge points may originate from the gray value contrast at the object surface, *e.g.* due to texture, inscription, shape discontinuities or figure/background separation. Small distortions in the imaging process and inaccuracies in determining the edge orientation are considered in equation (2.3) by parameter $\check{\delta}_2$, which specifies the upper bound for permissible errors.

Estimation of the Orientation of Gray Value Edges

In our system, the orientations of gray value edges are extracted by applying to the image a set of four differently oriented 2D *Gabor functions* and combining the responses appropriately [66, pp. 219-258].[4] The Gabor function is a Gauss-modulated, complex-harmonic function, which looks as follows (2D case).

$$f_{\mathcal{D},\mathcal{U}}^{Gb}(p) := exp\left(-\pi \cdot p^T \cdot (\mathcal{D})^{-2} \cdot p\right) \cdot exp\left(-\hat{i} \cdot 2 \cdot \pi \cdot \mathcal{U}^T \cdot p\right) \qquad (2.4)$$

The diagonal matrix $\mathcal{D} := diag(\sigma_1, \sigma_2)$ contains the eccentricity values of the Gaussian in two orthogonal directions, the vector $\mathcal{U} := (u_1, u_2)^T$ consists of the center frequencies, and $\hat{i}$ is the imaginary unit.

The general case of a rotated Gabor function is obtained by

$$f_{\psi,\mathcal{D},\mathcal{U}}^{Gb}(p) := f_{\mathcal{D},\mathcal{U}}^{Gb}(\mathcal{R}_\psi \cdot p), \quad with \ \ \mathcal{R}_\psi := \begin{pmatrix} \cos(\psi) & -\sin(\psi) \\ \sin(\psi) & \cos(\psi) \end{pmatrix} \qquad (2.5)$$

Four rotated versions of Gabor functions are defined for $\psi := 0°$, $45°$, $90°$, and $135°$, which means that the individual filters respond most sensitive to edges whose orientations (*i.e.* gradient angles) are equal to the rotation angle of the filter. The specific choice of filter orientations reveals considerable simplifications in successive computations.

The edge orientation is estimated from the amplitudes $\mathcal{I}_1^A(p)$, $\mathcal{I}_2^A(p)$, $\mathcal{I}_3^A(p)$, $\mathcal{I}_4^A(p)$ of the complex response of the four filters. These amplitudes are multiplied with the *cosine* of the doubled angle and added up, and this procedure is repeated with the *sine* of the doubled angles. From the two results, we compute the *arcus tangens* but take the quadrant of the coordinate system into account, transform the result into an angle which must be discretized within the integer set $\{0, \cdots, 179\}$, and do simple exception handling at singularities.

---

[4] A much simpler approach would combine the responses of just two orthogonally directed Sobel filters. However, our Gabor-based approach can be parameterized flexibly, and thus the orientation of gray value edges can be estimated more accurately.

$$\mathcal{I}_1^A(p) \cdot \cos(0°) + \mathcal{I}_2^A(p) \cdot \cos(90°) + \mathcal{I}_3^A(p) \cdot \cos(180°) + $$
$$\mathcal{I}_4^A(p) \cdot \cos(270°) = \quad \mathcal{I}_1^A(p) - \mathcal{I}_3^A(p) \tag{2.6}$$

$$\mathcal{I}_1^A(p) \cdot \sin(0°) + \mathcal{I}_2^A(p) \cdot \sin(90°) + \mathcal{I}_3^A(p) \cdot \sin(180°) + $$
$$\mathcal{I}_4^A(p) \cdot \sin(270°) = \quad \mathcal{I}_2^A(p) - \mathcal{I}_4^A(p) \tag{2.7}$$

$$\mathcal{I}_1^H(p) := -0.5 \cdot \arctan\left(\mathcal{I}_2^A(p) - \mathcal{I}_4^A(p), \mathcal{I}_1^A(p) - \mathcal{I}_3^A(p)\right) \tag{2.8}$$

$$\mathcal{I}_2^H(p) := \left\{ \begin{array}{lll} \pi + \mathcal{I}_1^H(p) & : & \mathcal{I}_1^H(p) < 0 \\ \mathcal{I}_1^H(p) & : & \mathcal{I}_1^H(p) \geq 0 \end{array} \right. \tag{2.9}$$

$$\mathcal{I}^O(p) := \text{round}\left((\mathcal{I}_2^H(p)/\pi) \cdot 180°\right) \tag{2.10}$$

The standard Hough transformation is modified by taking the orientation at each edge point into account and accumulating only those small set of parameter tuples, for which equation (2.3) holds. A tolerance band $\tilde{\delta}_2$ is introduced to take the inaccuracy of the edge orientation $\mathcal{I}^O(p)$ at position p into account. The parameter $\tilde{\delta}_2$ in equation (2.3) correlates to δ_2 in equation (2.11), therefore, in the following we refer only to δ_2.

Definition 2.3 (Orientation-selective Hough transformation, OHT)
The orientation-selective Hough transformation (OHT) of the binary image $\mathcal{I}^B$ and the orientation image $\mathcal{I}^O$ relative to the polar form f^L of a straight line is of functionality $\mathcal{Q} \to [0, \cdots, (I_w \cdot I_h)]$. The resulting Hough image $\mathcal{I}^{OH}$ is defined by

$$\mathcal{I}^{OH}(q) := \#\left\{p \in \mathcal{P} \mid \mathcal{I}^B(p) = 1 \wedge |f^L(p, q)| \leq \delta_1 \wedge \right.$$
$$\left. (\phi - \delta_2) \leq \mathcal{I}^O(p) \leq (\phi + \delta_2)\right\} \tag{2.11}$$

Figure 2.5 shows the resulting Hough image $\mathcal{I}^{OH}$ of the OHT if we assign $\delta_2 = 2°$ angle degrees for the tolerance band of edge orientations. Compared to the Hough image $\mathcal{I}^{SH}$ of the SHT in Figure 2.4 we realize that more local maxima are sharpened in the Hough image $\mathcal{I}^{OH}$ of the OHT.

The local maxima can be obtained iteratively by looking for the global maximum, erasing the peak position together with a small surrounding area, and restarting the search for the next maximum. Due to the sharpness of the peaks in $\mathcal{I}^{OH}$, it is much easier (compared to $\mathcal{I}^{SH}$) to control the area size to be erased in each iteration. Figure 2.6 shows the extracted lines specified by the set of 10 most maximal peaks in the Hough images of SHT and OHT (top and bottom, respectively). Obviously, the lines extracted with OHT, which consider the line/edge orientation compatibility, are more relevant and accurate for describing the object boundary. The line/edge orientation compatibility not only supports the extraction of relevant lines, but is also useful for verifying or adjusting a subset of candidate lines, which are determined in the context of other approaches (see Section 2.3 and Section 2.4 later on).

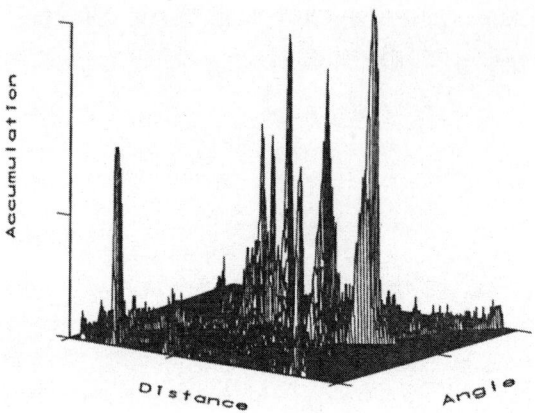

Fig. 2.5. Orientation-selective Hough transformation of the binarized image in Figure 2.3 by taking an image of edge orientations into account. Several local maxima are much more sharpened than in the Hough image of SHT in Figure 2.4.

Fig. 2.6. Extracted image lines based on 10 most maximal peaks in the Hough image of SHT (top) and of OHT (bottom). The lines extracted with OHT are more relevant and accurate for describing the object boundary.

Experiments to the Line/Edge Orientation Compatibility

The orientation compatibility is used to verify certain segments of lines, *i.e.* restrict the unbounded lines extracted with OHT to the relevant segments of an object boundary. The finite set of discrete points $p_i, i \in \{1, \cdots, N\}$, of a line bounded between p_1 and p_N is denoted by $\mathcal{L}(p_1, p_N)$. In Figure 2.7 the line segment $\mathcal{L}(p_a, p_d)$ through the characteristic points $\{p_a, p_b, p_c, p_d\}$ is only relevant between p_b and p_c. Figure 2.8 shows the course of edge orientation for the points p_i located on the line segment $\mathcal{L}(p_a, p_d)$. The horizontal axis is for the points on the line segment and the vertical axis for the orientations. Furthermore, the orientation ϕ of the line segment $\mathcal{L}(p_a, p_d)$ is depicted, which is of course independent of the points on the line. For the points of the line segment $\mathcal{L}(p_b, p_c)$ we obtain small deviation values between the edge orientations and the line orientation. On the other hand, there is large variance in the set of deviation values coming from edge orientations of the points of the line segments $\mathcal{L}(p_a, p_b)$ or $\mathcal{L}(p_c, p_d)$.

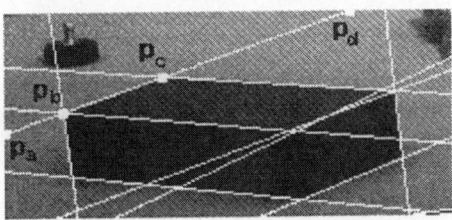

Fig. 2.7. Example line with characteristic points $\{p_a, p_b, p_c, p_d\}$, defined by intersection with other lines and with the image border. Just the line segment between p_b and p_c is relevant for the boundary.

For verifying a line segment, we evaluate the deviation between the orientation of the line and the orientations of the gray value edges of all points on the line segment.

Definition 2.4 (Orientation-deviation related to a line segment) *The orientation-deviation between orientation ϕ of a line and the orientations of all edges on a segment $\mathcal{L}(p_1, p_N)$ of the line is defined by*

$$D_{LE}(\phi, \mathcal{L}(p_1, p_N)) := \frac{1}{N} \cdot \sum_{i=1}^{N} D_{OO}(\phi, \mathcal{I}^O(p_i)) \qquad (2.12)$$

$$D_{OO}(\phi, \mathcal{I}^O(p_i)) := \frac{\min\{|d_i|, |d_i + 180°|, |d_i - 180°|\}}{90°} \qquad (2.13)$$

$$d_i := \phi - \mathcal{I}^O(p_i) \qquad (2.14)$$

The minimization involved in equation (2.13) is due to the restriction of edge and line orientation in the angle interval $[0°, \cdots, 180°]$, respectively. For

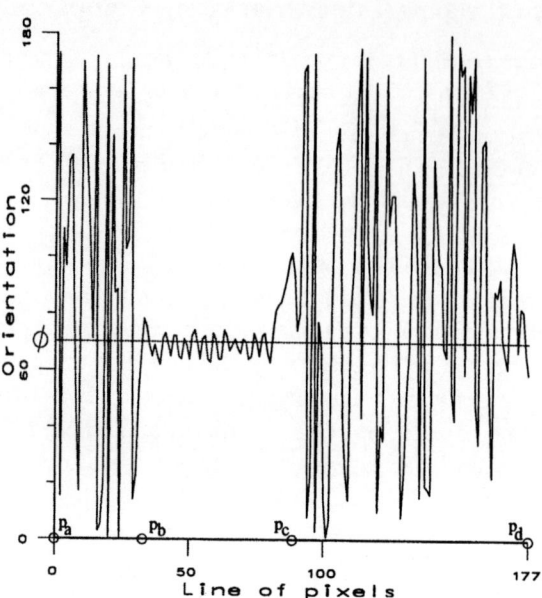

Fig. 2.8. Course of edge orientation of points along the line $\mathcal{L}(p_a, p_d)$, and orientation ϕ of this line. Small deviation within the relevant segment $\mathcal{L}(p_b, p_c)$, and large deviations within the other two segments.

example, the deviation between a line orientation $\phi = 0°$ and an edge orientation $\mathcal{I}^O(p_i) = 180°$ must be defined to be zero. Furthermore, a normalization factor is introduced to restrict the deviation values in the real unit interval. The orientation-deviation related to the line segments $\mathcal{L}(p_a, p_b)$, $\mathcal{L}(p_b, p_c)$, and $\mathcal{L}(p_c, p_d)$ in Figure 2.7 is shown in Figure 2.9. For line segment $\mathcal{L}(p_b, p_c)$ it is minimal, as expected, because this line segment originates from the actual boundary of the target object.

Based on the definition for orientation-deviation, we can formally introduce the line/edge orientation compatibility between the orientation of a line and the orientations of all edges on a segment of the line.

Assumption 2.2 (Line/edge orientation compatibility for a line segment, LEOC) *Let δ_3 be the permissible orientation-deviation in the sense of a necessary geometric/photometric compatibility. The line/edge orientation compatibility (LEOC) holds between the orientation ϕ of a line and the orientations $\mathcal{I}^O(p_i)$ of all edges on a segment $\mathcal{L}(p_1, p_N)$ of the line if*

$$D_{LE}(\phi, \mathcal{L}(p_1, p_N)) \leq \delta_3 \tag{2.15}$$

For example, Figure 2.9 shows that the line/edge orientation compatibility just holds for the line segment $\mathcal{L}(p_b, p_c)$ if we apply a compatibility threshold $\delta_3 = 0.15$. For the extraction of object boundaries, it is desirable to have

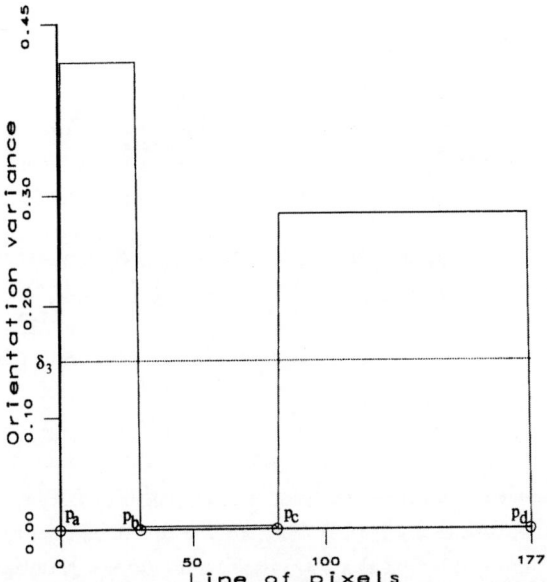

Fig. 2.9. Mean variance of the edge orientations for three line segments $\mathcal{L}(p_a, p_b)$, $\mathcal{L}(p_b, p_c)$, $\mathcal{L}(p_c, p_d)$ related to line orientation ϕ.

threshold δ_3 specified such that the line/edge orientation compatibility just holds for the line segments which are relevant for a boundary.

Appropriate values for thresholds δ_1 and δ_2 in Definition 2.3, and δ_3 in Assumption 2.2 must be determined on the basis of visual demonstration (see Section 2.5). Furthermore, the parameters $\mathcal{D}$ (Gaussian eccentricity values) and $\mathcal{U}$ (center frequencies) of the Gabor function in equation (2.4) are determined by task-relevant experimentation.

2.2.3 Junction Compatibility between Pencils and Corners

The geometric *line* feature and the photometric *edge* feature are one-dimensional in nature. A further sophisticated compatibility criterion can be defined on the basis of two-dimensional image structures. In the projected object boundary, usually, two or more lines meet at a common point (see points p_b and p_c in Figure 2.7). A collection of non-parallel image line segments meeting at a common point is designated as *pencil of lines*, and the common point is designated as *pencil point* [52, pp. 17-18].

At the pencil point a *gray value corner* should be detected in the image. Generally, gray value corners are located at the curvature extrema along edge sequences. A review of several corner detectors was presented by Rohr [140], however, we used the recently published *SUSAN operator* [159]. Exemplary,

Figure 2.10 shows a set of gray value corners with the corners at the points p_b and p_c included.

Fig. 2.10. The SUSAN operator has detected a set of gray value corners shown as black squares. We find all those gray value corners included which are characteristic for the three-dimensional object boundary.

According to this, we must consider a compatibility between the geometric *pencil* feature and the photometric *corner* feature. Common attributes are needed for characterizing a junction of lines and a junction of edge sequences. The term *junction* is used as generic term both for the geometric pencil feature and the photometric corner feature. We define an *M-junction* to consist of M meeting lines, or consist of M meeting edge sequences, respectively. An *M-junction of lines* can be characterized by the position p_{pc} of the pencil point and the orientations $\mathcal{A} := (\alpha_1, \cdots, \alpha_M)$ of the meeting lines related to the horizontal axis. Similary, an *M-junction of edge sequences* is characterized by the position p_{cr} of the corner point and the orientations $\mathcal{B} := (\beta_1, \cdots, \beta_M)$ of the meeting edge sequences against the horizontal axis.

Definition 2.5 (Junction-deviation related to a pencil) *The junction-deviation between an M-junction of lines with orientations $\mathcal{A}$ at the pencil point p_{pc} and an M-junction of edge sequences with orientations $\mathcal{B}$ at the corner point p_{cr} is defined by*

$$D_{PC}(p_{pc}, p_{cr}, \mathcal{A}, \mathcal{B}) := \omega_1 \cdot D_{JP}(p_{pc}, p_{cr}) + \omega_2 \cdot D_{JO}(\mathcal{A}, \mathcal{B}) \qquad (2.16)$$

$$D_{JP}(p_{pc}, p_{cr}) := \frac{\|p_{pc} - p_{cr}\|}{I_d} \qquad (2.17)$$

$$D_{JO}(\mathcal{A}, \mathcal{B}) := \frac{1}{180° \cdot M} \cdot \sum_{i=1}^{M} \min\{|d_i|, |d_i + 360°|, |d_i - 360°|\} \qquad (2.18)$$

$$d_i := \alpha_i - \beta_i \qquad (2.19)$$

Equation (2.16) combines two components of junction-deviation with the factors ω_1 and ω_2, which are used to weight each part. The first component (*i.e.* equation (2.17)) evaluates the euclidean distance between pencil and corner point. The second component (*i.e.* equation (2.18)) computes the deviation between the orientation of a line and of the corresponding edge sequence, and this is done for all corresponding pairs in order to compute a mean value. Both components and the final outcome of equation (2.16) are normalized in the real unit interval. Based on this definition, we formally introduce a pencil/corner junction compatibility.

Assumption 2.3 (Pencil/corner junction compatibility, PCJC) *Let δ_4 be the permissible junction-deviation in the sense of a necessary geometric/photometric compatibility. The pencil/corner junction compatibility (PCJC) holds between an M-junction of lines and an M-junction of edge sequences if*

$$D_{PC}(p_{pc}, p_{cr}, \mathcal{A}, \mathcal{B}) \le \delta_4 \qquad (2.20)$$

This pencil/corner junction compatibility is used as a criterion to evaluate whether a junction belongs to the boundary of a target object. For illustration, the criterion is applied to three junctions designated in Figure 2.11 by the indices $1, 2, 3$. It is remarkable that junctions 1 and 2 belong to the boundary of the target object but junction 3 does not. The white squares show the pencil points $p_{pc1}, p_{pc2}, p_{pc3}$. They are determined by intersection of straight lines extracted via OHT (see also Figure 2.6 (bottom)). The black squares show the corner points $p_{cr1}, p_{cr2}, p_{cr3}$. They have been selected from the whole set of corner points in Figure 2.10 based on nearest neighborhood to the pencil points. Related to equation (2.17), we realize that all three pencil points have corresponding corner points in close neighborhood. For the three junctions $1, 2, 3$ we obtain position deviations (normalized by the image diagonal I_d) of about $0.009, 0.006, 0.008$, respectively. According to this, a small position deviation is only a necessary but not a sufficient criterion to classify junction 3 unlike to 1 and 2. Therefore, the junctions must be characterized in more detail, which is considered in equation (2.18).

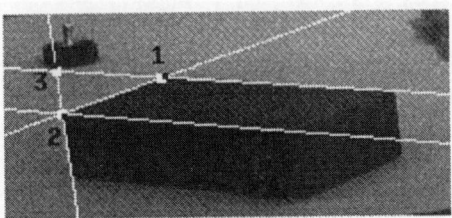

Fig. 2.11. A subset of three gray value corners is selected. They are located in close neighborhood to three pencil points of relevant boundary lines, respectively. Just the corners at junctions 1 and 2 are relevant for boundary extraction but not the corner at junction 3.

Characterization of Gray Value Corners

A *steerable wedge filter*, adopted from Simoncelli and Farid [158], has been applied at the pencil points $p_{pc1}, p_{pc2}, p_{pc3}$ in order to locally characterize the gray value structure. We prefer pencil points instead of the neighboring corner points, because the pencil points arise from line detection, which is more robust than corner detection. A wedge is rotating in discrete steps around a pencil point and at each step the mean gray value within the wedge mask is computed. For example, the wedge started in horizontal direction pointing to the right and than rotated counter-clockwise in increments of $4°$ angle degrees. According to the steerability property of this filter, we compute the filter response at the basic orientations, and approximate filter responses in between two basic orientations (if necessary). This gives a one-dimensional course of smoothed gray values around the pencil point. The first derivative of a one-dimensional Gaussian is applied to this course and the magnitude is computed from it. We obtain for each discrete orientation a *significance measurement* for the existence of an edge sequence having just this orientation and starting at the pencil point. As a result, a curve of filter responses is obtained which characterizes the gray value structure around the pencil point.

Experiments to the Pencil/Corner Junction Compatibility

These curves are shown for the junctions 1, 2, and 3 in Figure 2.12, Figure 2.13, and Figure 2.14, respectively. The curve in Figure 2.12 shows two local maxima (near to $0°$ and $360°$, respectively, and near to $200°$) indicating a 2-junction. The curve in Figure 2.13 shows three local maxima (near to $20°$, $290°$, and $360°$) indicating a 3-junction. The curve in Figure 2.14 shows two local maxima (near to $20°$ and $150°$) indicating a 2-junction. The vertical dotted lines in each figure indicate the orientation of the image lines (extracted by Hough transformation), which converge at the three junctions, respectively. We clearly observe that in Figure 2.12 and Figure 2.13 the local maxima are located near to the orientations of the converging lines. However, in Figure 2.14 the positions of the curve maxima and of the dotted lines differ significantly as junction 3 does not belong to the object boundary. By applying the formula in equation (2.18), we compute about $0.04, 0.03, 0.83$ for the three junctions $1, 2, 3$, respectively. Based on a threshold, we can easily conclude that the pencil/corner junction compatibility holds for junctions 1 and 2, but not for junction 3. Appropriate values for δ_4 in Assumption 2.3 and the parameters of the SUSAN corner detector are determined on the basis of visual demonstration.

The line/edge orientation compatibility and the pencil/corner junction compatibility can be assumed generally for all scenes containing approximate polyhedral objects. In Sections 2.3 and 2.4, the evaluation of the junction compatibility according to equation (2.16) is combined with the evaluation of the orientation compatibility according to equation (2.12). This measure will be used in combination with regularity features of object shapes to define the relevance of certain line segments for the boundary description.

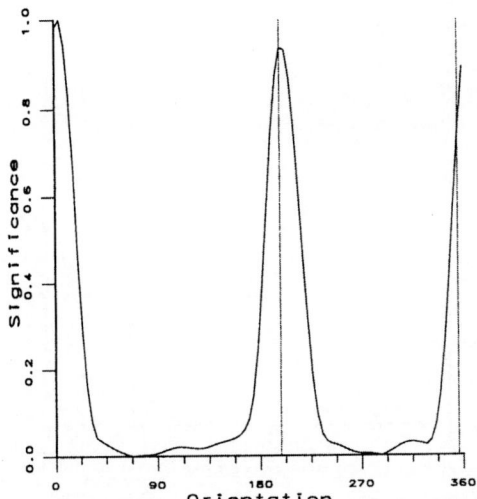

Fig. 2.12. Response of a counter-clockwise rotating wedge filter applied at junction 1. The two local maxima indicate a 2-junction, *i.e.* two converging edge sequences. The orientations of the converging edge sequences (maxima of the curve) are similar to the orientations of two converging lines (denoted by the positions of the vertical lines). The pencil/corner junction compatibility holds for junction 1.

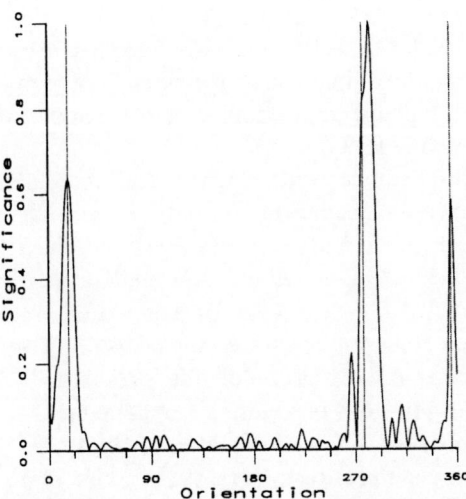

Fig. 2.13. Response of a counter-clockwise rotating wedge filter applied at junction 2. The three local maxima indicate a 3-junction, *i.e.* three converging edge sequences. The maxima of the curve are located near the positions of the vertical lines. The pencil/corner junction compatibility holds for junction 2.

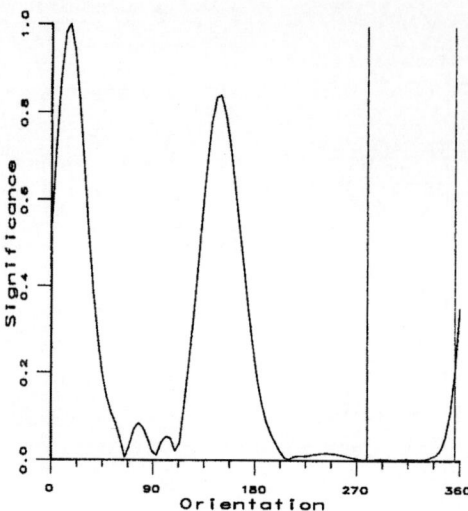

Fig. 2.14. Response of a counter-clockwise rotating wedge filter applied at junction 3. The two local maxima indicate a 2-junction. The orientations of the converging edge sequences are dissimilar to those of the converging lines. The pencil/corner junction compatibility does not hold for junction 3.

2.3 Compatibility-Based Structural Level Grouping

The orientation-selective Hough transformation (OHT), the line/edge orientation compatibility (LEOC) and the pencil/corner junction compatibility (PCJC) are the basis for detecting high-level geometric structures in the image. Additionally, we introduce another principle of geometric/photometric compatibility, *i.e.* the phase compatibility between approximate parallel lines and gray value ramps (PRPC).

These compatibilities between geometric and photometric image features are combined with pure geometric compatibilities under projective transformation. The geometric compatibilities are determined for geometric regularity features which are inherent in man-made 3D objects. Approximate parallel or right-angled line segments, or approximate reflection-symmetric or translation-symmetric polylines are considered in a sophisticated search strategy for detecting organizations of line segments. This section focuses on the extraction of polygons originating from the faces or the silhouettes of approximate polyhedral 3D objects. Related to the aspect of extracting polygons, our approach is similar to a work of Havaldar *et al.* [77], who extract closed figures, *e.g.* by discovering approximate reflection-symmetric polylines (they call it *skewed symmetries between super segments*). The principal distinction to our work is that we are striving for an integration of grouping cues from the structural level with cues from other levels, *i.e.* from signal level and assembly level.

2.3.1 Hough Peaks for Approximate Parallel Lines

It is well-known for the projective transformation of an ideal pinhole camera that an image point $(x_1, x_2)^T$ is computed from a 3D scene point $(y_1, y_2, y_3)^T$ by

$$x_1 := b \cdot \frac{y_1}{y_3} \quad , \quad x_2 := b \cdot \frac{y_2}{y_3} \tag{2.21}$$

Parameter b is the distance between the lens center and the projection plane. According to equation (2.21), it is obvious that parallel 3D lines are no longer parallel after projective transformation to the image (except for lines parallel to the projection plane).

Fortunately, for certain imaging conditions the parallelism is almost invariant under projective transformation. In order to obtain an impression for this, we describe the imaging condition for taking the picture in Figure 2.15. The distance between the camera and the target object was about $1000mm$, and the lens of the objective was of $12mm$ focal length. In this case the deviation from parallelism, which depends on object orientation, is at most $6°$ angle degrees. We formulate the *parallelism compatibility* and relate it to the configuration of Hough peaks.

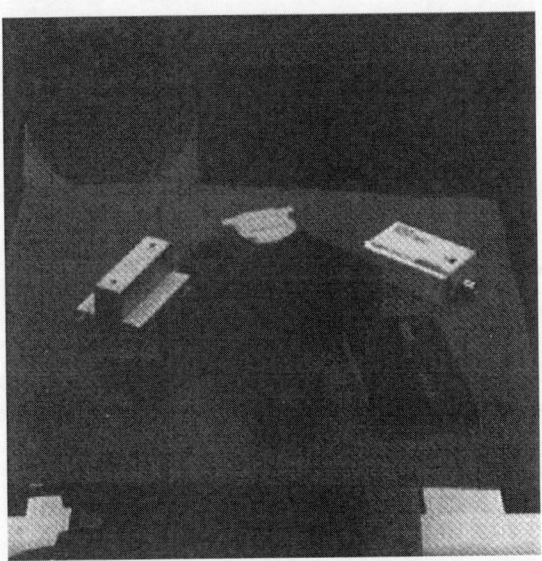

Fig. 2.15. Arrangement of objects in a scene of electrical scrap. The black dummy box is our target object for the purpose of demonstration. It is located in a complex environment, is partially occluded, and has a protrusing socket.

Definition 2.6 (Approximate parallel lines) *Let δ_5 be the permissible deviation from parallelism, i.e. maximal deviation from exact regularity. Two* *image lines with values ϕ_1 and ϕ_2 of the angle parameter are approximate parallel if*

$$D_O(\phi_1, \phi_2) \leq \delta_5 \tag{2.22}$$

Assumption 2.4 (Parallelism compatibility) *The parallelism compatibility holds if parallel lines in 3D are approximate parallel after projective transformation. For such imaging conditions, parallel lines in 3D occur as peaks in the Hough image being located within a horizontal stripe of height δ_5.*

The peaks in a *Hough stripe* describe approximate parallel lines in the gray value image. Figure 2.16 shows the Hough image obtained from the gray value image in Figure 2.15 after binarization (edge detection) and application of the OHT.

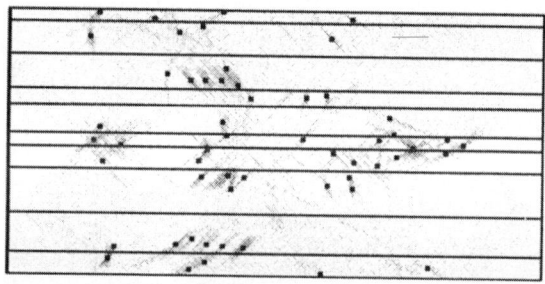

Fig. 2.16. Hough image obtained after edge detection in the image of Figure 2.15. A set 55 most maximal peaks is marked by black dots. They have been organized in 10 clusters (horizontal stripes) using the ISODATA clustering algorithm.

The Hough image has been edited with black squares and horizontal lines, which mark a set of 55 local maxima organized in 10 horizontal stripes. The local maxima are obtained using the approach mentioned in Subsection 2.2.2. For grouping the peak positions, we solely take parameter ϕ into account and use the distance function D_O from equation (2.13). According to this, angles near to $0°$ can be grouped with angles near to $180°$. A procedure similar to the error-based *ISODATA clustering algorithm* can be applied but taking the modified distance function into account [149, pp. 109-125]. Initially, the algorithm groups vectors (in this application, simply scalars) by using the standard *K–means* method. Then, clusters exhibiting large variances are split in two, and clusters that are too close together are merged. Next, K–means is reiterated taking the new clusters into account. This sequence is repeated until no more clusters are split or merged. The merging/splitting parameters are taken in agreement with the pre-specified δ_5 from Assumption 2.4.

For example, Figure 2.17 shows the set of approximate parallel lines specified by the Hough peaks in the fourth stripe, and Figure 2.18 shows it for the eighth stripe of Hough peaks. Under these lines we find candidates for describing the boundary of the dummy box.

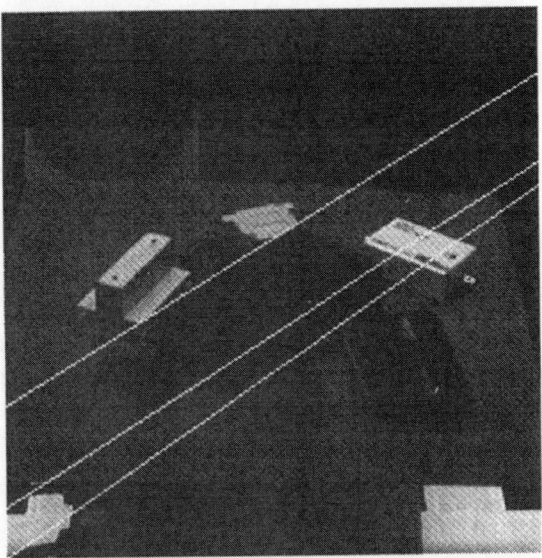

Fig. 2.17. The set of approximate parallel lines specified by the Hough peaks in the fourth stripe of Figure 2.16.

The next subsection introduces a criterion for grouping approximate parallel lines which are supposed to belong to the silhouette of an object in the image. It is the *phase compatibility* between approximate parallel lines and gray value ramps (PRPC).

2.3.2 Phase Compatibility between Parallels and Ramps

In a Fourier-transformed image the phase plays a much greater role for exhibiting the relevant image structure than the amplitude [85]. It is a global image feature in the sense of taking the whole image into account. Instead, the *local phase* is an image feature for characterizing image structures locally [66, pp. 258-278]. In addition to describing a gray value edge by the gradient angle and magnitude, it is interesting to distinguish various types of edges, *e.g.* roof and ramp image structures. For example, the boundary edges of a homogeneous surface can be classified as ramps, and the edges of inscription strokes are more of the roof type. A further distinction can be made whether the ramp is from left to right or vice versa, and whether the roof is directed to top or bottom. The four special cases and all intermediate situations can be arranged on a unit cycle in the complex plane and the distinction is with the polar angle φ. The specific values $\varphi := 0°$ or $\varphi := 180°$ represent the

Fig. 2.18. The set of approximate parallel lines specified by the Hough peaks in the eighth stripe of Figure 2.16.

two roof types, and $\varphi := 90°$ or $\varphi := 270°$ represent the two ramp types (see Figure 2.19).

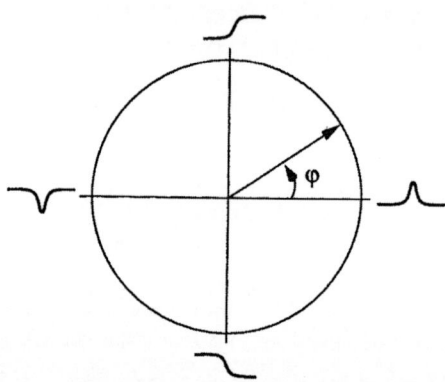

Fig. 2.19. Representation of local phase as a vector in the complex plane where the argument φ reflects the roof and ramp relationship (evenness and oddness) of edge characterization.

Local Phase Computation with Gabor Functions

For characterizing the type of gray value edges we can apply once more the Gabor function, which has already been used in Section 2.2 for estimating edge orientation. The Gabor function is a so-called *analytic function*, in which the imaginary part is the *Hilbert transform* of the real part. The real part is even and is tuned to respond on roof edges, and the imaginary part is odd and is tuned to respond on ramp edges. The local phase of a Gabor-transformed image actually reveals the edge type at a certain point, *i.e.* the polar angle φ. However, the local phase is a one-dimensional construct which should be determined along the direction of the gradient angle (edge orientation). According to this, for characterizing an edge one must first apply four rotated Gabor functions ($0°, 45°, 90°, 135°$) and determine from the local amplitudes the edge orientation, and second apply another Gabor function in the direction of the estimated edge orientation and determine from the local phase the edge type.

Experiments to Local Phase Computation

For example, let us assume a simulated image with a region R_1 of homogeneous gray value g_1 and an environmental region R_2 of homogeneous gray value g_2. If the gradient angles (in the interval $[0°, \cdots, 360°]$) at the boundary points of R_1 are considered for computing the local phases in the relevant directions, then a constant local phase value $\frac{\pi}{2}$ reveals for any point of the boundary. However, we restricted the computation of edge orientation to the interval $[0°, \cdots, 180°]$, and in consequence of this, the local phase takes on either $\frac{\pi}{2}$ or $-\frac{\pi}{2}$. Specifically, if region R_1 is a parallelogram then the local phases for pairs of points taken from opposite parallel lines have different signs.

The change of sign in the local phase can also be observed for pairs of approximate parallel boundary lines from the silhouette of the dummy box. Figure 2.20 illustrates two similar cases in the left and right column. The top picture shows two parallel boundary lines which are exemplary taken from the picture in Figure 2.6 (bottom). In the middle two short lines are depicted which cross the boundary lines and are approximate orthogonal to them. The local phases are computed (based on edge orientation) for the set of points from these short lines instead of just the one boundary point. This is to obtain an impression for the sensitivity of local phase computation in the neighborhood of the relevant point.[5] The bottom diagram shows the two courses of local phases computed for the two sets of line points from the middle picture. The local phases of one line are located in the interval $[0, \cdots, \pi]$ and of the other line in the interval $[-\pi, \cdots, 0]$.

---

[5] Log-normal filters are proposed to treat this problem [66, pp. 219-288].

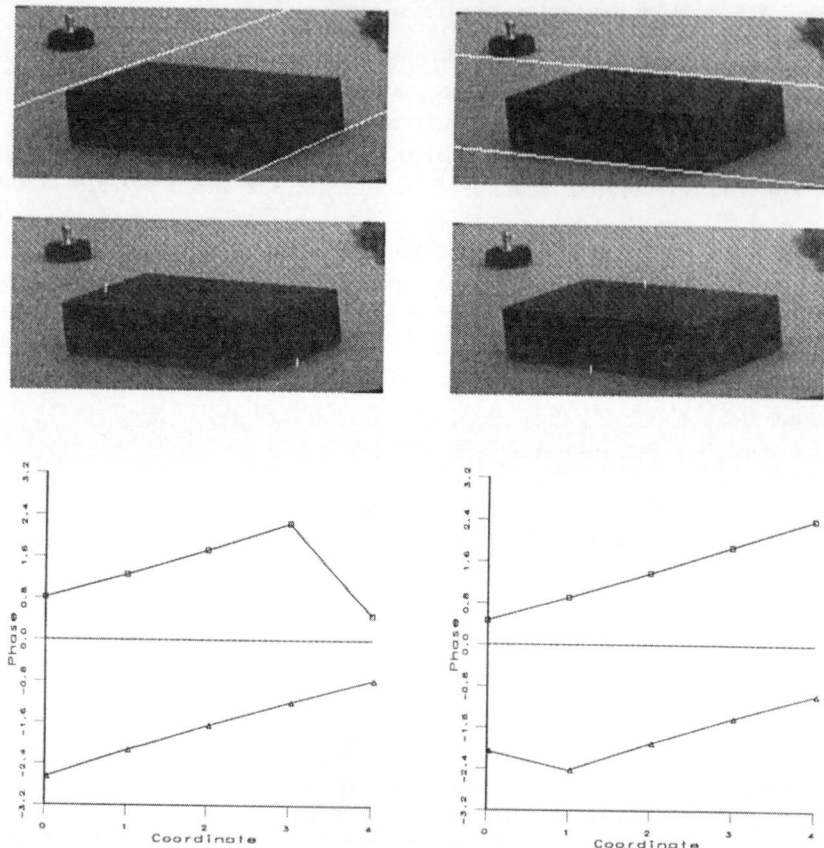

Fig. 2.20. Dummy box, two pairs of parallel lines, phase computation.

In Figure 2.21, the picture at the top contains again the dummy box, an artificial vertical line segment through the box region, and three dots where the line segment crosses three boundary lines (the boundary lines are not depicted). The diagram at the bottom shows the course of local phases for all points of the line segment from top to bottom. Furthermore, the mentioned crossing points in the picture are indicated in the diagram at specific positions of the coordinate axis (see positions of vertical lines in the diagram). The local phase at the first and second crossing point is negative, and at the third point positive, and the values are approximately $-\frac{\pi}{2}$ and $+\frac{\pi}{2}$, respectively. Beyond these specific points, *i.e.* in nearly homogeneous image regions, the local phases are approximately 0.

Local Phase Computations at Opposite Boundary Lines

Based on this observation, it is possible to formulate a necessary criterion for grouping silhouette lines. Two approximate parallel line segments belong to

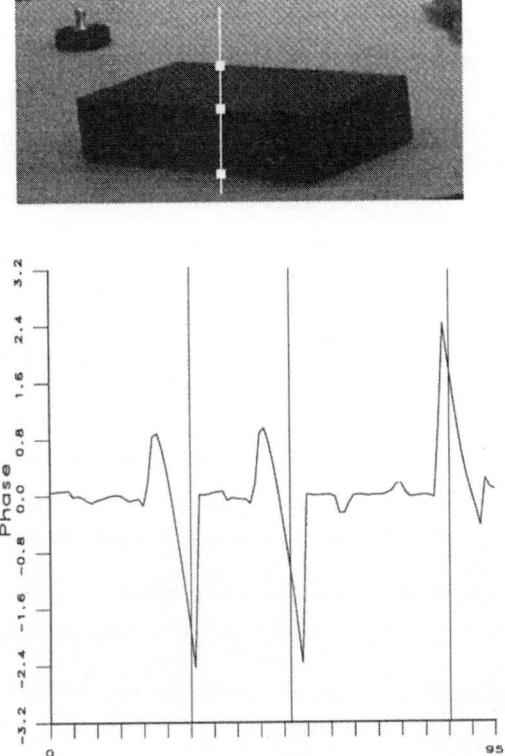

Fig. 2.21. Dummy box, vertical line and crossing points, course of local phases.

the silhouette of an object if gray value ramps at the first line segment are converse to the gray value ramps at the second line segment. It is provided that local phases are determined along the direction of the gradient of gray value edges. Furthermore, the gray values of the object must be generally higher or generally lower than the gray values of the background. Let $\mathcal{L}_1$ and $\mathcal{L}_2$ be two approximate parallel line segments. For all points of $\mathcal{L}_1$ we compute the local phases and take the mean, designated by function application $f^{ph}(\mathcal{L}_1)$, and the same procedure is repeated for $\mathcal{L}_2$, which is designated by $f^{ph}(\mathcal{L}_2)$. Then, we define a similarity measure between phases such that the similarity between equal phases is 1, and the similarity between phases with opposite directions in the complex plane is 0 (see Figure 2.19).

$$D_{PR}(\mathcal{L}_1, \mathcal{L}_2) := \left| 1 - \frac{|f^{ph}(\mathcal{L}_1) - f^{ph}(\mathcal{L}_2)|}{\pi} \right| \tag{2.23}$$

The usefulness of equation (2.23) can be illustrated by various examples of line segment pairs in a simulated image.[6] The image may consist of a black rectangle R_1 and a white environment R_2. For two line segments taken from opposite sides of R_1 the mean phases are $-\frac{\pi}{2}$ and $+\frac{\pi}{2}$, respectively, and therefore $D_{PR} = 0$. For two line segments taken from one side of R_1 the mean phases are equal, i.e. $-\frac{\pi}{2}$ or $+\frac{\pi}{2}$, and therefore $D_{PR} = 1$. For two line segments taken from the interior of region R_1 the mean phases are 0, respectively, and therefore again $D_{PR} = 1$. If one segment is taken from one side of R_1, i.e. mean phase is $-\frac{\pi}{2}$ or $+\frac{\pi}{2}$, and the other line segment taken from the interior of region R_1, i.e. mean phase is 0, then $D_{PR} = 0.5$. According to this discussion, for line segments taken from opposite sides of a silhouette the measure reveals 0, and in the other cases the measure is larger than 0. Generally, the value of measure D_{PR} is restricted in the unit interval. Based on this definition, we can formally introduce a phase compatibility between approximate parallel lines and gray value ramps.

Assumption 2.5 (Parallel/ramp phase compatibility, PRPC) *It is assumed that two line segments $\mathcal{L}_1$ and $\mathcal{L}_2$ are approximately parallel according to equation (2.22). Furthermore, the gray values between the two segments should be generally higher or alternatively lower than beyond the segments. Let δ_6 be a threshold for the necessary parallel/ramp phase compatibility. The parallel/ramp phase compatibility (PRPC) holds between a pair of approximate parallel lines if*

$$D_{PR}(\mathcal{L}_1, \mathcal{L}_2) \leq \delta_6 \tag{2.24}$$

Appropriate task-relevant thresholds δ_5 and δ_6 can be determined on the basis of visual demonstration (see Section 2.5).[7]

In the next sections, the horizontal clusters of Hough peaks are used in combination with the LEOC, PCJC, and PRPC principles for extracting the faces or silhouettes of objects (or merely approximate faces or silhouettes) from the images.

2.3.3 Extraction of Regular Quadrangles

In a multitude of man-made objects the faces or the silhouettes can be approximated by squares, rectangles, or trapezoids. The projective transformation of these shapes yields approximations of *squares, rhombuses, rectangles, parallelograms,* or *trapezoids.* A generic procedure will be presented for extracting from the image these specific quadrangles.[8] Under the constraint of

---

[6] The extraction of line segments from lines in real images is treated afterwards.

[7] In Section 2.5, we present an approach of local phase estimation which becomes more robust if phase computation is extended to a small environment near the line segment. The respective consideration of environmental points is for reducing the sensitivity of local phase computation (see above).

[8] The next subsection, after this one, presents a procedure for extracting more general polygons.

clustered Hough peaks, we exhaustively look for quadruples of Hough peaks, extract the four line segments by line intersection, respectively, apply the LEOC, PCJC, and PRPC principles, and determine deviations from a certain standard form. For each quadrangle all evaluations are combined, which results in a saliency value including both photometric and geometric aspects.

Geometric/Photometric Compatibility for Quadrangles

We have introduced the orientation-deviation D_{LE} related to a line segment in equation (2.12) and the junction-deviation D_{PC} related to a pencil of line segments in equation (2.16). To extend the LEOC and PCJC principles to quadrangles, we simply average these values for the four segments and for the four pencils, respectively.

$$D_{LE\_QD} := \frac{1}{4} \cdot \sum_{i=1}^{4} D_{LE\_i}, \quad D_{PC\_QD} := \frac{1}{4} \cdot \sum_{i=1}^{4} D_{PC\_i} \qquad (2.25)$$

For convenience, we omitted the parameters and simply introduced an index for the line segments involved in a quadrangle. The resulting functions D_{LE_QD} and D_{PC_QD} can be used in combination to define a *geometric/photometric compatibility for quadrangles*.

Assumption 2.6 (Geometric/photometric compatibility for a quadrangle) *The necessary geometric/photometric compatibility for a quadrangle is specified by parameter δ_7. The geometric/photometric compatibility for a quadrangle holds, if*

$$(D_{LE\_QD} + D_{PC\_QD}) \leq \delta_7 \qquad (2.26)$$

For specific quadrangle shapes with one or two pairs of approximate parallel lines, the PRPC principle can be included in Assumption 2.6. The left hand side of equation (2.26) is extended by a further summation term D_{PR_QD} which describes the parallel/ramp phase compatibility related to a specific quadrangle having approximate parallel lines.

Geometric Deviation of Quadrangles from Standard Forms

In order to consider the pure geometric aspect, we define the deviation of a quadrangle from certain standard forms. For the sequence of four line segments of a quadrangle, let $\mathcal{H} := (l_1^s, l_2^s, l_3^s, l_4^s)$ be the lengths, $\mathcal{G} := (\gamma_1, \gamma_2, \gamma_3, \gamma_4)$ be the inner angles of two successive segments, and $\mathcal{F} := (\phi_1, \phi_2, \phi_3, \phi_4)$ be the orientation angles of the polar form representations.

Definition 2.7 (Rectangle-deviation, parallelogram-deviation, square-deviation, rhombus-deviation, trapezoid-deviation)

The rectangle-deviation of a quadrangle is defined by

$$D_{RC}(\mathcal{G}) := \frac{1}{4 \cdot 360°} \cdot \sum_{i=1}^{4} |\gamma_i - 90°| \tag{2.27}$$

The parallelogram-deviation of a quadrangle is defined by

$$D_{PA}(\mathcal{G}) := \frac{1}{2 \cdot 360°} \cdot (|\gamma_1 - \gamma_3| + |\gamma_2 - \gamma_4|) \tag{2.28}$$

The square-deviation of a quadrangle is defined by

$$D_{SQ}(\mathcal{G}, \mathcal{H}) := \frac{1}{2} \cdot (D_{RC}(\mathcal{G}) + V_{SL}(\mathcal{H})) \tag{2.29}$$

with the normalized length variance $V_{SL}(\mathcal{H})$ of the four line segments.

The rhombus-deviation of a quadrangle is defined by

$$D_{RH}(\mathcal{G}, \mathcal{H}) := \frac{1}{2} \cdot (D_{PA}(\mathcal{G}) + V_{SL}(\mathcal{H})) \tag{2.30}$$

The trapezoid-deviation of a quadrangle is defined by

$$D_{TR}(\mathcal{F}) := \min\{D_O(\phi_1, \phi_3), D_O(\phi_2, \phi_4)\} \tag{2.31}$$

Normalization factors are chosen such that the possible values of each function fall in the unit interval, respectively. For the rectangle-deviation the mean deviation from right-angles is computed for the inner angles of the quadrangle. For the parallelogram-deviation the mean difference between diagonally opposite inner angles is computed. The square-deviation and rhombus-deviation are based on the former definitions and additionally include the variance of the lengths of line segments. The trapezoid-deviation is based on equation (2.13) and computes for the two pairs of diagonally opposite line segments the minimum of deviation from parallelism.

The features related to the geometric/photometric compatibility for quadrangles, and the feature related to the geometric deviation of quadrangles from specific shapes must be combined to give measures of conspicuity of certain shapes in an image.

Definition 2.8 (Saliency of specific quadrangles) *The saliency of a specific quadrangle is defined by*

$$A_{SP\_QD} := 1 - \frac{D_{QD}}{\omega_1 + \omega_2 + \omega_3 + \omega_4}, \quad with \tag{2.32}$$

$$D_{QD} := \omega_1 \cdot D_{LE\_QD} + \omega_2 \cdot D_{PC\_QD} + \\ \omega_3 \cdot D_{PR\_QD} + \omega_4 \cdot D_{SP\_QD} \tag{2.33}$$

The specific quadrangle can be an approximate rectangle, parallelogram, square, rhombus, or trapezoid, and for these cases the generic function symbol D_{SP_QD} must be replaced by D_{RC}, D_{PA}, D_{SQ}, D_{RH}, or D_{TR}, as introduced in Definition 2.7.

Generic Procedure PE_1 for the Extraction of Specific Quadrangles

---

Procedure PE_1

---

1. For each pair of cluster stripes in the set of Hough peaks:
1.1. For each pair of Hough peaks in the first stripe:
1.1.1. For each pair of Hough peaks in the second stripe:
1.1.1.1. Intersect the lines specified by the four Hough peaks and construct the quadrangle.
1.1.1.2. Compute the mean line/edge orientation-deviation using function D_{LE_QD}.
1.1.1.3. Compute the mean pencil/corner junction-deviation using function D_{PC_QD}.
1.1.1.4. Compute the mean parallel/ramp phase-deviation using function D_{PR_QD}.
1.1.1.5. Compute the deviation from the specific quadrangle using function D_{SP_QD}.
1.1.1.6. Compute the saliency value by combining the above results according to equation (2.32).
2. Bring the specific quadrangles into order according to decreasing saliency values.

The generic procedure works for all types of specific quadrangles which have been mentioned above, except for trapezoids. For the extraction of trapezoids the algorithm can be modified, such that it iterates over single cluster stripes, and selects all combinations of two Hough peaks in each stripe, respectively, and takes the third and fourth Hough peaks from any other cluster stripes. Thus, it is considered that a trapezoid just consists of one pair of parallel line segments, instead of two such pairs of the other specific quadrangles.

Experiments to the Extraction of Regular Quadrangles

These procedures have been applied to complicated scenes of electrical scrap in order to draw conclusions concerning the usefulness of the principles introduced above. The goal of the following experiments was to extract from the images specific quadrangles which describe the faces or silhouettes of objects.

First, we applied the procedure to the image in Figure 2.15 with the intention of extracting approximate parallelograms. In the saliency measure, the weighting factors $\omega_1, \omega_2, \omega_3, \omega_4$ are set equal to 0.25. Figure 2.22 shows exemplary a set of 65 approximate parallelograms, which are best according to the saliency measure. The decreasing course of saliency values for the approximate parallelograms is depicted in Figure 2.23. The silhouette boundary of the dummy box (see Figure 2.24) is included as number three. For automatically detecting the dummy box, *i.e.* determining the reference number three as the relevant one, it is necessary to apply object recognition. In Chapter 3, we present approaches for object recognition, which can be applied within the areas of a certain set of most salient quadrangles.

Second, the procedure for boundary extraction is used to extract approximate rectangles with the intention of extracting the electronic board of a computer interior in Figure 2.25. The best set of 10 approximate rectangles are outlined in black color, including the relevant one of the board.

Third, the procedure for boundary extraction was applied to extract approximate rhombuses with the goal of locating the electronic board in the image of Figure 2.26. The best set of 3 approximate rhombuses are outlined in white color, including the relevant one.

Finally, the extraction of approximate trapezoids is shown in Figure 2.27.

Fig. 2.22. Based on a saliency measure a subset of most conspicuous, approximate parallelogramms have been extracted.

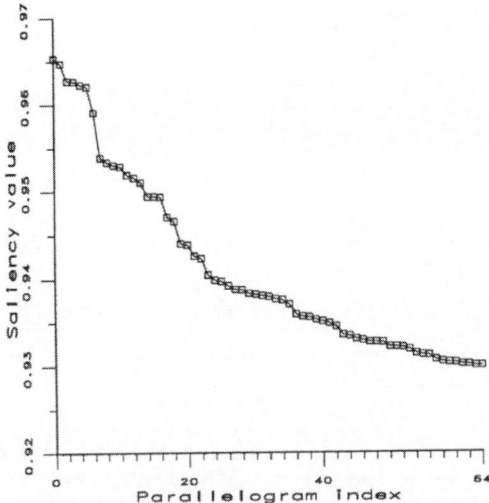

Fig. 2.23. Decreasing course of saliency values for the subset of extracted parallelograms in the image of Figure 2.22.

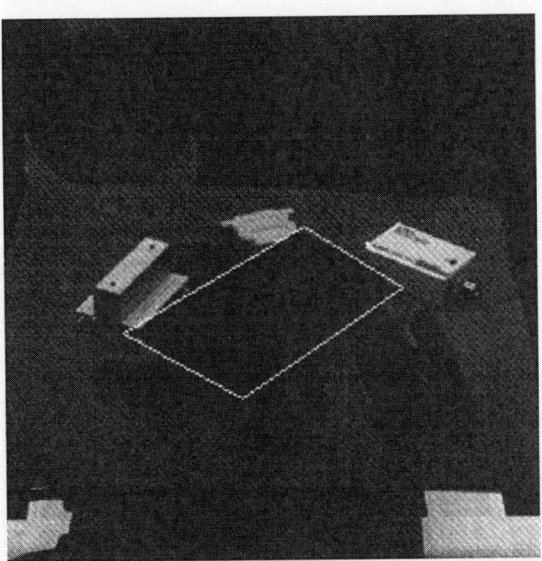

Fig. 2.24. The silhouette boundary of the dummy box is included in the set of most conspicuous, approximate parallelogramms.

Fig. 2.25. Electronic board of a computer interior and an extracted subset of approximate rectangles. One of these rectangles represents the boundary of the board.

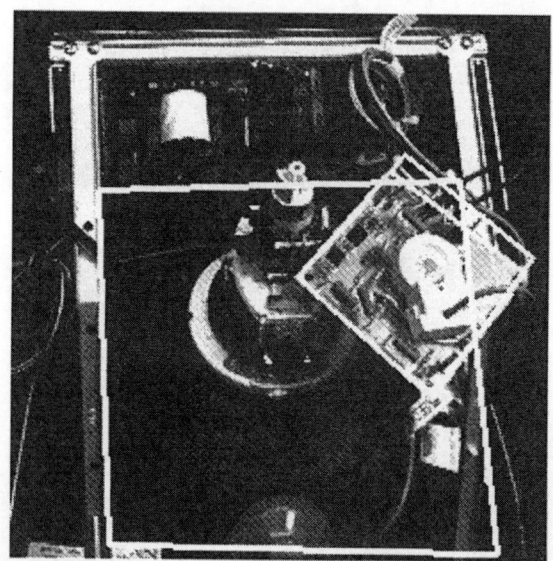

Fig. 2.26. Computer interior containing an electronic board. One of the extracted approximate rhombuses represents the boundary of the board.

Fig. 2.27. Image of a loudspeaker and a subset of extracted approximate trapezoids. One of these represents a side-face of the loudspeaker.

2.3.4 Extraction of Regular Polygons

The measures D_{LE} and D_{PC} involved in the geometric/photometric compatibility of single line segments can easily be extended to polygons of K line segments.

$$D_{LE\_PG} := \frac{1}{K} \cdot \sum_{i=1}^{K} D_{LE\_i} , \quad D_{PC\_PG} := \frac{1}{K} \cdot \sum_{i=1}^{K} D_{PC\_i} \qquad (2.34)$$

Number K is set a priori which may be known from the task sepcification.

Assumption 2.7 (Geometric/photometric compatibility for a polygon) *The necessary geometric/ photometric compatibility for a polygon is specified by parameter δ_8. The geometric/photometric compatibility for a polygon holds if*

$$(D_{LE\_PG} + D_{PC\_PG}) \le \delta_8 \qquad (2.35)$$

These compatibility features must be combined with pure geometric features of the polygon. By considering specific polygon regularities, a measurement of conspicuity is obtained, *i.e.* a saliency value for the polygon. More complex regularities are interesting than the simple ones involved in specific quadrangles. For polygons with arbitrary segment number, we define three types of regularities, which are general in the sense that they typically appear in scenes of man-made objects. Two types among the regularities are based on symmetries between polylines, *i.e.* the reflection-symmetry and the translation-symmetry. A reflection-symmetry is a pair of polylines in which each one can be obtained by reflecting the other one at an axis. A translation-symmetry is a pair of polylines in which each one can be obtained by reflecting the other one sequentially at two parallel axes (Figure 2.28). Approximate reflection-symmetric and translation-symmetric polylines appear exemplary in the image of a computer monitor (see in Figure 2.29 the two polygons outlined with white color).

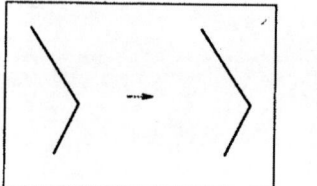 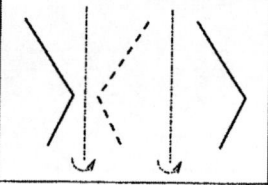

Fig. 2.28. Constructing a translation of polylines by a two-step reflection at two parallel axes.

Fig. 2.29. Computer monitor with two types of regularities of the polygonal faces. The side face is approximate reflection-symmetric, and the top face is approximate translation-symmetric.

The third type of regularity is a right-angle, *i.e.* two successive line segments of a polygon are right-angled, respectively. For example, Figure 2.30 shows an electronic board with approximate right-angled shape.

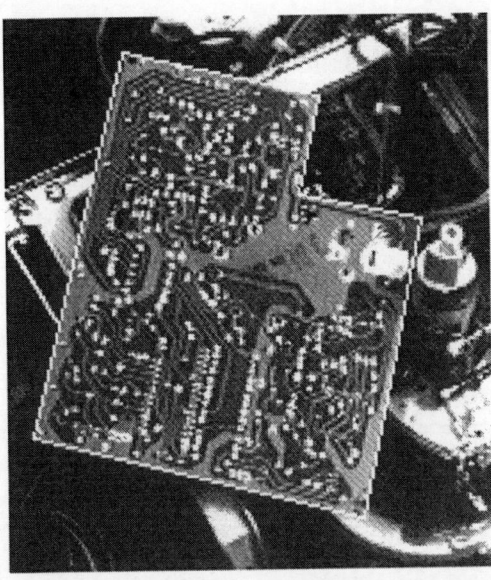

Fig. 2.30. Computer interior with an electronic board, and a hexagonal boundary. The pencils of lines of the board hexagon are approximate right-angles.

Regularities of Polygons

The basic component for describing polygon regularities is a *polyline*. It is a non-closed and non-branching sequence of connected line segments. We construct for each polygon a pair of non-overlapping polylines with equal numbers of line segments. A polygon with an odd number of line segments is the union of two polylines and one single line segment (see in Figure 2.31). For a polygon with an even number of line segments we distinguish two cases. First, the polygon can be the union of two polylines, *i.e.* they meet at two polygon junctions (see in Figure 2.31, middle). Second, the polygon can be the union of two polylines and of two single line segments located at the end of each polyline, respectively (see Figure 2.31, bottom).

Let $\mathcal{G} := (\gamma_1, \cdots, \gamma_K)$ be the ordered sequence of inner angles of the polygon. A candidate pair of polylines is represented by $\mathcal{G}^1 := (\gamma_1^1, \cdots, \gamma_k^1)$ and $\mathcal{G}^2 := (\gamma_1^2, \cdots, \gamma_k^2)$, *i.e.* the sequence of inner angles related to the first polyline and the opposite (corresponding) inner angles related to the second polyline. All angles γ_i^1 or γ_i^2 are contained in $\mathcal{G}$. Included in $\mathcal{G}^1$ and $\mathcal{G}^2$ are the inner angles at the end of the polylines, where one polyline meets the other one or meets a single line segment. There are different candidate pairs

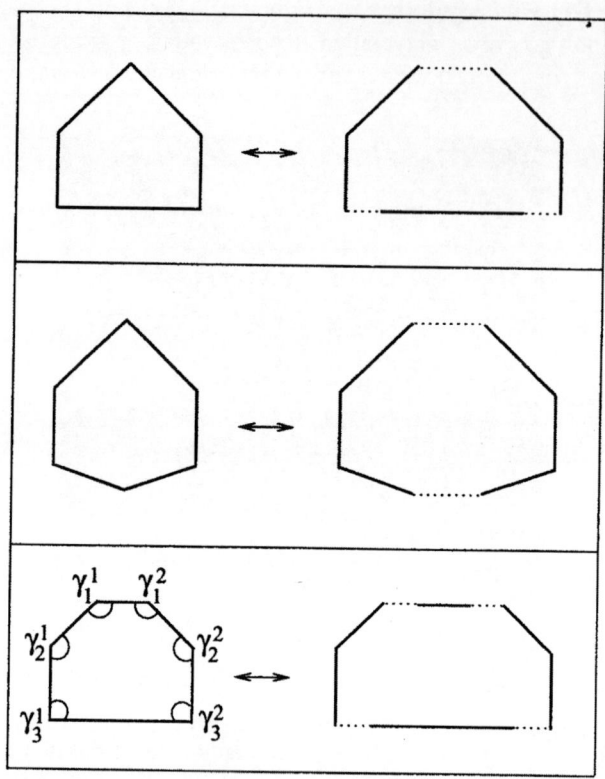

Fig. 2.31. Organization of reflection-symmetric polygons by two polylines with equal number of line segments and up to two single line segments.

of polylines of a polygon, *i.e.* the pair $(\mathcal{G}^1, \mathcal{G}^2)$ of angle sequences is just one element in a whole set which we designate by $\mathcal{G}_{\dot{\Sigma}}$. For example, Figure 2.31 (bottom) shows for one candidate pair of polylines the two corresponding sequences of inner polygon angles.

Definition 2.9 (Reflection-symmetric polygon) *A polygon is reflection-symmetric if a pair of polylines exists with sequences $\mathcal{G}^1$ and $\mathcal{G}^2$ of inner angles such that $d_{rs}(\gamma_i^1, \gamma_i^2) = 0°$ for each tuple (γ_i^1, γ_i^2), $i \in \{1, \cdots, k\}$,*

$$d_{rs}(\gamma_i^1, \gamma_i^2) := |\gamma_i^1 - \gamma_i^2| \tag{2.36}$$

Figure 2.31 shows three reflection-symmetric polygons (left column) and the relevant configuration of polylines and single line segments (right column). Obviously, for all three polygons there exist a vertical axis of reflection for mapping one polyline onto the other. Examplary, in the bottom polygon the following equations hold,

$$\gamma_1^1 = \gamma_1^2, \quad \gamma_2^1 = \gamma_2^2, \quad \gamma_3^1 = \gamma_3^2 \tag{2.37}$$

Definition 2.10 (Translation-symmetric polygon) *A polygon is translation-symmetric if a pair of polylines exists with sequences $\mathcal{G}^1$ and $\mathcal{G}^2$ of inner angles such that $d_{ts}(\gamma_i^1, \gamma_i^2) = 0°$ for each tuple $(\gamma_i^1, \gamma_i^2), i \in \{1, \cdots, k\}$,*

$$d_{ts}(\gamma_i^1, \gamma_i^2) := \left\{ \begin{array}{ll} |\gamma_i^1 - (360° - \gamma_i^2)| & : \quad i \in \{2, \cdots, k-1\} \\ |\gamma_i^1 - (180° - \gamma_i^2)| & : \quad i \in \{1, k\} \end{array} \right. \qquad (2.38)$$

For the translation-symmetry, it is plausible to match the inner polygon angles (γ_i^1) of the first polyline with the corresponding exterior angles $(360° - \gamma_i^2)$ of the second polyline. The corresponding angles at the end of the two polylines, respectively must be matched modulo 180° angle degrees. Figure 2.32 shows a hexagon with translation-symmetry and the relevant pair of polylines. We can imagine a translation vector for mapping one polyline onto the other. The property of translation-symmetry is easily realized by

$$\gamma_1^1 = (180° - \gamma_1^2), \quad \gamma_2^1 = (360° - \gamma_2^2), \quad \gamma_3^1 = (180° - \gamma_3^2) \qquad (2.39)$$

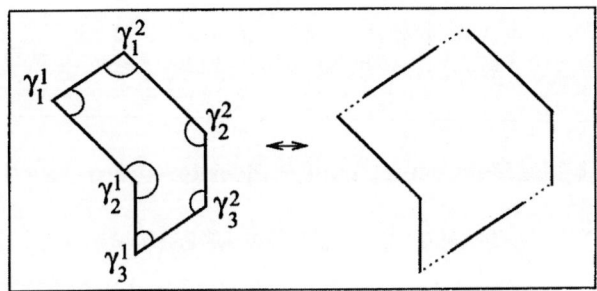

Fig. 2.32. Organization of a translation-symmetric hexagon by two polylines and two single line segments.

For a polygon the property of reflection-symmetry or translation-symmetry is examined by determining whether there is a pair of polylines for which the proposition in Definitions 2.9 and 2.10 holds, respectively. For this purpose, one has to evaluate all possible pairs of polylines by applying a parallel or sequential algorithm. If there is no relevant pair, then the polygon is not regular concerning reflection- or translation-symmetry. Figure 2.33 shows two polygons with inappropriate pairs of polylines, although the reflection-symmetry and the translation-symmetry respectively has already been realized in Figure 2.31 (bottom) and Figure 2.32.

For a polygon the deviation from reflection-symmetry and translation-symmetry respectively is defined by matching for all candidate pairs of polylines the two sequences of inner angles and taking the minimum value.

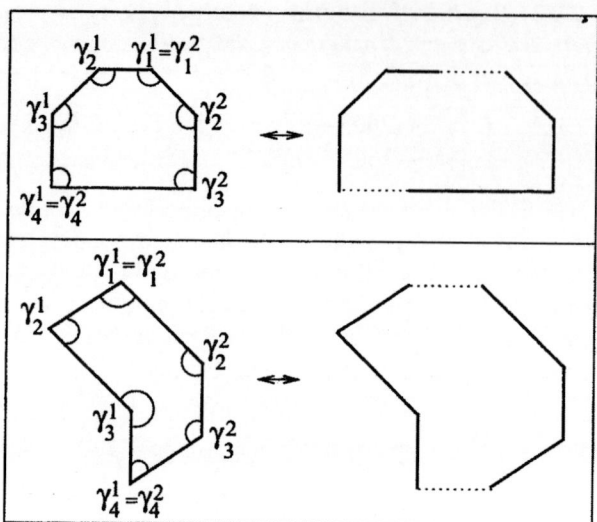

Fig. 2.33. Polygons with inappropriate organizations of pairs of polylines. The reflection-symmetry and the translation-symmetry have been verified based on appropriate organizations in Figure 2.31 (bottom) and Figure 2.32, respectively.

Definition 2.11 (Deviation from reflection-symmetry or translation-symmetry)

For a polygon the deviation from reflection-symmetry is

$$D_{RS} := \min_{\mathcal{G}_\Sigma} \left\{ \frac{1}{k \cdot 360°} \cdot \sum_{i=1}^{k} d_{rs}(\gamma_i^1, \gamma_i^2) \right\} \qquad (2.40)$$

For a polygon the deviation from translation-symmetry is

$$D_{TS} := \min_{\mathcal{G}_\Sigma} \left\{ \frac{1}{k \cdot 360°} \cdot \sum_{i=1}^{k} d_{ts}(\gamma_i^1, \gamma_i^2) \right\} \qquad (2.41)$$

For the reflection-symmetric polygons in Figure 2.31 the equation $D_{RS} = 0$ is obtained, and for the translation-symmetric polygon in Figure 2.32 the equation $D_{TS} = 0$. However, for the left face of the computer monitor in Figure 2.29 we compute $D_{RS} \approx 0.064$, which indicates an approximate reflection-symmetry. For the top face of the computer monitor we compute $D_{TS} \approx 0.015$, which indicates an approximate translation-symmetry.

Definition 2.12 (Approximate reflection-symmetric or approximate translation-symmetric polygons) *Let δ_9 and δ_{10} be the permissible deviations from exact reflection- and translation-symmetry, respectively. A polygon is approximate reflection-symmetric if $D_{RS} \leq \delta_9$. A polygon is approximate translation-symmetric if $D_{TS} \leq \delta_{10}$.*

Finally, we consider the right-angled polygon as a third type of typical regularity in man-made objects. In the special case of convex right-angled polygons, the shape is a rectangle. In general, the polygon can include concavities with the inner polygon angle 270°.

Definition 2.13 (Right-angled polygon, right-angle deviation) *A polygon is right-angled if for every pencil of two line segments the inner polygon angle is given by* $d_{ra}(\gamma_i) = 0°$*, with*

$$d_{ra}(\gamma_i) := \min\{|\gamma_i - 90°|, |\gamma_i - 270°|\} \tag{2.42}$$

The right-angle deviation of a polygon with M pencils of two line segments repectively, i.e. a polygon with M line segments, is defined by

$$D_{RA}(\mathcal{G}) := \frac{1}{M \cdot 360°} \cdot \sum_{i=1}^{M} d_{ra}(\gamma_i) \tag{2.43}$$

Definition 2.14 (Approximate right-angled polygon) *Let δ_{11} be the permissible deviation from right-angled polygons. A polygon is approximate right-angled if $D_{RA} \leq \delta_{11}$.*

Based on these definitions, we introduce three compatibilities under projective transformation. For the imaging conditions in our experiments a threshold value $\delta_9 = \delta_{10} = \delta_{11} = 0.1$ proved as appropriate. The compatibilities will be used later on for extracting regular polygons from the image.

Assumption 2.8 (Reflection-symmetry compatibility) *The reflection-symmetry compatibility for a polygon holds, if a reflection-symmetric 3D polygon is approximate reflection-symmetric after projective transformation.*

Assumption 2.9 (Translation-symmetry compatibility) *The translation-symmetry compatibility for a polygon holds if a translation-symmetric 3D polygon is approximate translation-symmetric after projective transformation.*

Assumption 2.10 (Right-angle compatibility) *The right-angle compatibility for a polygon holds if a right-angled 3D polygon is approximate right-angled after projective transformation.*

The projective transformation of 3D object faces, which are supposed to be reflection-symmetric, translation-symmetric or right-angled polygons, yields approximations of these specific polygons in the image. The features related to the geometric deviation of polygons from these specific shapes must be combined with features related to the geometric/photometric compatibility for polygons. This gives a measure of conspicuity of specific polygons in an image.

Definition 2.15 (Saliency of specific polygons) *The saliency of a specific polygon is defined by*

$$A_{SP\_PG} := 1 - \frac{D_{PG}}{\omega_1 + \omega_2 + \omega_3 + \omega_4}, \quad with \tag{2.44}$$

$$D_{PG} := \omega_1 \cdot D_{LE\_PG} + \omega_2 \cdot D_{PC\_PG} +$$
$$\omega_3 \cdot D_{PR\_PG} + \omega_4 \cdot D_{SP\_PG} \tag{2.45}$$

The function symbol D_{SP_PG} must be replaced by D_{RS}, D_{TS}, or D_{RA}, depending on whether there is interest in approximate reflection-symmetric, translation-symmetric or right-angled polygons.

Especially, in the case of reflection-symmetric polygons it makes sense to apply the principle of parallel/ramp phase compatibility, which is included in equation (2.44) by the term D_{PR_PG}. For other shapes the relevant parameter ω_3 is set to 0.

Generic Procedure PE_2 for the Extraction of Specific Polygons

---

Procedure PE_2

---

1. From the whole set of combinations of three Hough peaks:
1.1. Select just the combinations under the constraint that first and third Hough peak don't belong to the same cluster as the second peak.
1.2. Determine for each combination a line segment by intersecting first and third line with the second one (specified by the Hough peaks, respectively).
1.3. Select the line segments, which are completely contained in the image, and are not isolated.
1.4. Compute the line/edge orientation-deviation using function D_{LE}, and the pencil/corner junction-deviation using function D_{PC}, and select those line segments, for which both the LEOC and PCJC principles hold.
2. Compute a graph representing the neighborhood of line segments, *i.e.* create a knot for each intersection point and an arc for each line segment.
3. Compute the set of minimal, planar cycles in the graph, *i.e.* minimal numbers of knots and no arc in the graph is intersecting the cycles. This gives a candidate set of polygons representing faces of an object.

Procedure PE_2, continued

---

4. For each polygon:
   4.1. Compute the mean line/edge orientation-deviation using function D_{LE_PG}.
   4.2. Compute the mean pencil/corner junction-deviation using function D_{PC_PG}.
   4.3. Compute the mean parallel/ramp phase-deviation using function D_{PR_PG}.
   4.4. Compute the deviation from a specific regularity using generic function D_{SP_PG}.
   4.5. Compute the saliency value by combining the above results according to equation (2.44).
5. Bring the specific polygons into order according to decreasing saliency values.

Experiments to the Extraction of Regular Polygons

This generic procedure has been applied successfully for localizing regular polygons which originate from the surfaces of man-made objects. For example, the side and top face of the computer monitor in Figure 2.29 have been extracted. They were determined most saliently as approximate reflection-symmetric and approximate translation-symmetric octagons, respectively. As a second example, the boundary of the electronic board in Figure 2.30 has been extracted. It was determined most saliently as approximate right-angled hexagon.

Further examples are presented in the next section in the framework of extracting arrangements of polygons. For example, the complete arrangement of polygons for the computer monitor will be determined by extracting and slightly adjusting the polygons of the side, top, and front faces under the consideration of certain assembly level constraints.

2.4 Compatibility-Based Assembly Level Grouping

The extraction of regular polygons can be considered as an intermediate step of the higher goal of localizing certain objects and describing their boundaries in more detail. Geometric/photometric compatible features have been combined with geometric regularity features for defining a saliency measure of specific polygons in the image. A salient polygon may arise from the boundary of a single object face or of a whole object silhouette. In general, it is assumed that the surface of a man-made 3D object can be subdivided in several faces and from these only a subset is supposed to be observable, *e.g.*

in the case of a parallelepiped just 3 plane faces are observable from a non-degenerate view point. The projective transformation of this kind of object surface should yield an arrangement of several polygons. We introduce two assembly level grouping criteria, *i.e.* the *vanishing-point compatibility* and the *pencil compatibility*, which impose imperative restrictions on the shape of an arrangement of polygons. These constraints are directly correlated to the three-dimensional (regular) nature of the object surface. The principles are demonstrated for objects of roughly polyhedral shape, *i.e.* local protrusion, local deepening, or round corners are accepted.

2.4.1 Focusing Image Processing on Polygonal Windows

The compatibilities and regularities, used so far, just take basic principles of image formation and qualitative aspects of the shape of man-made objects into account. Although only general assumptions are involved, various experiments have shown that the extracted polygons are a useful basis for applying techniques of detailed object detection and boundary extraction. For example, Figure 2.30 showed an electronic board which has been extracted in a cluttered environment as an approximate right-angled hexagon. Subsequent image processing can focus on the hexagon image window for detecting specific electronic components on the board. As another example, Figure 2.22 showed objects of electrical scrap and a set of extracted approximate parallelograms, among which the rough silhouette boundary of the dummy box was included (see Figure 2.24). Subsequent image processing can focus on this parallelogram image window for extracting a detailed boundary description.

Polygons for the Approximation of Depicted Object Silhouettes

This section concentrates on detailed boundary extraction, and in this context, the previously extracted polygons serve a further purpose. We need to examine the type of geometric shape which is bounded by a polygon, in order to apply a relevant approach for *detailed* boundary extraction. The task of roughly characterizing the object shape (*e.g.* polyhedral or curvilinear) can be solved by taking a further constraint into account. According to the principles underlying the procedure of extracting polygons (in Section 2.3), it is reasonable to assume that there are polygons included, which approximately describe the *silhouette* of interesting objects. This has also been confirmed by the polygons extracted from the images in Figure 2.24 and Figure 2.30.

Assumption 2.11 (Silhouette approximation by salient polygons)
For the set of interesting objects, depicted in an image, there are salient polygons which approximate the silhouettes with a necessary accuracy δ_{12}.

Object Recognition in Polygonal Image Windows

Based on this assumption, it is expected that a large part of the polygon closely touches the object silhouette. Therefore, the gray value structure of the interior of a polygon belongs mainly to the appearance of *one* object and can be taken into account in various approaches of object or shape classification. For example, in some applications a simple histogram-based approach is appropriate, as was inspired by the work of Swain and Ballard [164]. An object is represented by taking several views from it, and computing histograms of gray values, edge orientations, corner properties, cooccurrence features, or further filter responses. In an offline phase a set of objects with relevant shapes is processed and the histograms stored in a database. In the online phase histograms are computed from the interior of a polygon, and matched with the database histograms. Based on the criterion of highest matching score the type of object shape must be determined in order to apply the relevant approach of boundary extraction, *e.g.* extraction of arrangements of polygons or alternatively curvilinear shapes. For example, from the gray value structure in the quadrangle image window in Figure 2.24 it has been concluded that the extraction of a detailed arrangement of polygons is reasonable. We have mentioned the approach of classification (which uses histograms) just briefly. For more complicated applications the simple approach is unsufficient and an advanced approach for classification is needed, *e.g.* see Matas *et al.* [107]. A detailed treatment is beyond the scope of this chapter.

Windowed Orientation-Selective Hough Transformation

The Figure 2.24 serves to demonstrate the principles underlying a generic procedure for extracting arrangements of polygons. Although the dummy object is located in a cluttered scene, the silhouette quadrangle is acting as a window and the procedure of boundary extraction is hardly detracted from the environment or background of the object. According to this, we introduce the *windowed orientation-selective Hough transformation (WOHT)* which just considers the image in a polygonal window. The definition is quite similar to that of OHT (see Definition 2.3), except that the votes are only collected from a subset $\mathcal{P}_S$ of coordinate tuples, taken from the interior of the extracted polygon and extended by a small band at the border. The WOHT contributes to overcome the problem of confusing profusion of Hough peaks. For example, we can apply the WOHT to the quadrangle window outlined in Figure 2.24, which contains the approximate parallelepiped object. The boundary line configuration for three visible faces should consist of nine line segments to be organized in three sets of three approximate parallel lines, respectively. In the Hough image of WOHT nine peaks must be organized in three stripes with three peaks in it, respectively. This constraint has to be considered in an approach for searching the relevant Hough peaks. However, a configuration like this is just a necessary characteristic but not a sufficient one for constructing the relevant object boundary. Further principles and compatibilities of projective transformation will be considered for the purpose of extracting the relevant arrangement of polygons.

Looking for Configurations of Hough Peaks

A basic procedure for extracting configurations of Hough peaks has already been mentioned in Subsection 2.3.1. It extracts a certain number of Hough peaks and groups them by considering only the line parameter ϕ. Related to the characteristic of parallelepiped objects the procedure must yield at least three clusters each consisting of at least three Hough peaks, respectively. Another alternative procedure can be designed which executes the search more carefully. It is looking for the global maximum peak and thus determines the first relevant horizontal stripe. Within the stripe a certain number of other maximum peaks must be localized. Then, the stripe is erased completely and in this modified Hough image the next global maximum is looked for. The new maximum defines the second relevant stripe in which once again the specified number of other maximum peaks are detected. By repeating the procedure a certain number of times we obtain the final configuration of Hough peaks. For demonstration, this procedure has been applied to the window in Figure 2.24. A configuration of nine Hough peaks organized in three stripes of three peaks respectively yields the set of image lines in Figure 2.34.

Fig. 2.34. Candidate set of nine boundary lines (for the dummy box) organized in three sets of three approximate parallel lines, respectively. Result of applying the windowed OHT to the quadrangle image window in Figure 2.24 and selecting nine Hough peaks organized in three stripes.

Although the necessary characteristic of the peak configuration holds, it is impossible to construct the complete object boundary, because important boundary lines are missing. Fortunately, a configuration of 12 Hough peaks organized in four stripes (see Figure 2.35) yields a more complete list of relevant boundary lines (see Figure 2.36). The next two subsections take compatibility principles and assembly level grouping criteria into account for evaluating or adjusting image lines for object boundary construction.

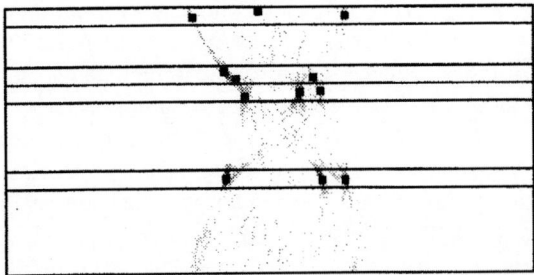

Fig. 2.35. Result of applying the windowed OHT to the quadrangle image window in Figure 2.24 and selecting 12 Hough peaks organized in four stripes of three Hough peaks, respectively.

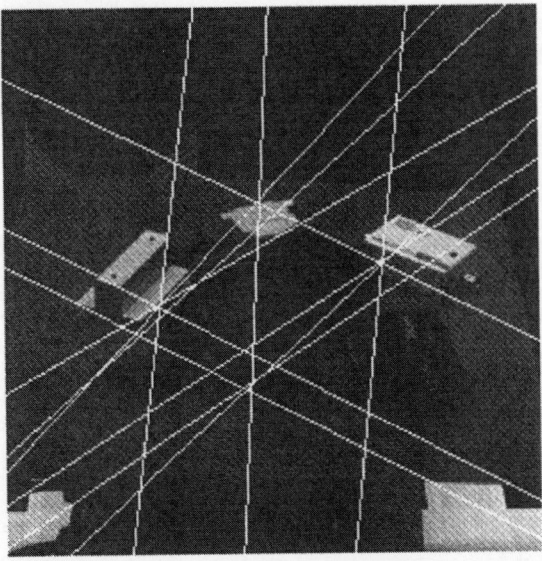

Fig. 2.36. Candidate set of 12 boundary lines (for the dummy box) specified by the 12 Hough peaks in Figure 2.35. More relevant relevant boundary lines are included compared to Figure 2.34.

2.4.2 Vanishing-Point Compatibility of Parallel Lines.

The projective transformation of parallel boundary lines generates approximate parallel image lines with the specific constraint that they should meet in one vanishing-point p_v. This vanishing-point compatibility imposes certain qualitative constraints on the courses of Hough peaks within a horizontal stripe (specifying approximate parallel lines). Figure 2.37 shows a projected parallelepiped and two vanishing-points p_{v1} and p_{v2} for two sets $\{\mathcal{L}_{11}, \mathcal{L}_{12}, \mathcal{L}_{13}\}$ and $\{\mathcal{L}_{21}, \mathcal{L}_{22}, \mathcal{L}_{23}\}$ of three approximate parallel line segments, respectively. Let (r_{ij}, ϕ_{ij}) be the polar form parameters of the lines, respectively. We realize for the monotonously increasing distance parameter $r_{11} < r_{12} < r_{13}$ of the first set of lines a monotonously *increasing* angle parameter $\phi_{11} < \phi_{12} < \phi_{13}$, and for the monotonously increasing distance parameter $r_{21} < r_{22} < r_{23}$ of the second set of lines a monotonously *decreasing* angle parameter $\phi_{21} > \phi_{22} > \phi_{23}$. This specific observation is generalized to the following geometric compatibility.

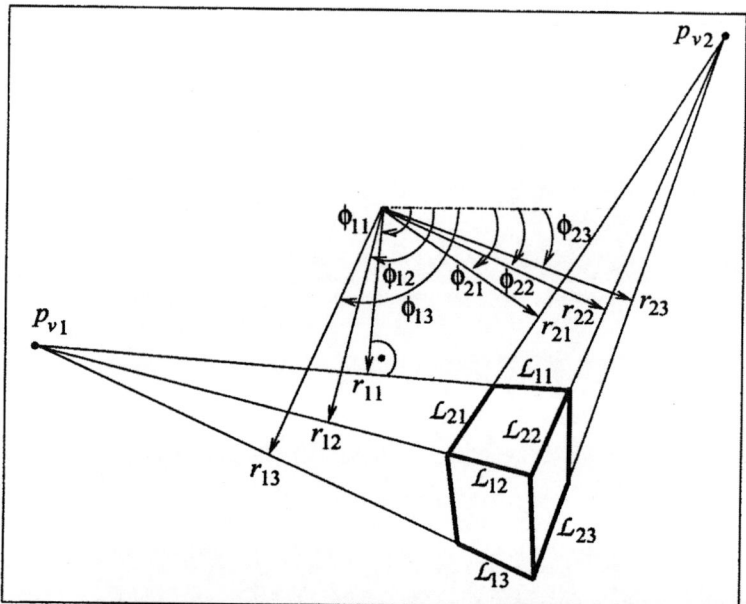

Fig. 2.37. Projected parallelepiped and two vanishing points p_{v1} and p_{v2}. Monotonously increasing angle parameter $\phi_{11} < \phi_{12} < \phi_{13}$, and monotonously decreasing angle parameter $\phi_{21} > \phi_{22} > \phi_{23}$ for two sets of three approximate parallel line segments, respectively.

Assumption 2.12 (Vanishing-point compatibility) *Let* $\{\mathcal{L}_1, \cdots, \mathcal{L}_V\}$ *be a set of approximate parallel line segments in the image, which originate from projective transformation of parallel line segments of the 3D object boundary. The extended lines related to the image line segments meet at a common vanishing point p_v and can be ordered according to the strong monotony $r_1 < \cdots < r_i < \cdots < r_V$ of the distance parameter. For this arrangement there is a weak monotony of the angle parameter,*

$$\phi_1 \geq \cdots \phi_i \geq \cdots \geq \phi_V \quad or \quad \phi_1 \leq \cdots \phi_i \leq \cdots \leq \phi_V \qquad (2.46)$$

Special Cases of the Vanishing-Point Compatibility

We have to be careful with approximate vertical lines whose angle parameter ϕ is near to $0°$ or near to $180°$. In a cluster of Hough peaks with that characterization all lines with ϕ near to $0°$ will be redefined by: $\hat{r} := -r$, and $\hat{\phi} := \phi + 180°$. This is permitted, because the equations $L(p, (r, \phi)) = 0$ and $L(p, (\hat{r}, \hat{\phi})) = 0$ define the same line (which is easily proven based on the definition for function L in equation (2.1)). Under this consideration the Assumption 2.12 must hold for any set of approximate parallel lines meeting at a common point. Consequently, the *course of Hough peaks* in a horizontal stripe must *increase* or *decrease* weak monotonously.

Experiments to the Vanishing-Point Compatibility

For demonstration, this vanishing-point compatibility will be examined in the Hough image of clustered peaks in Figure 2.35. Assumption 2.12 holds for the third and fourth stripe but not for the first and second stripe. Actually, the Hough peaks in the first stripe specify lines which are candidates for the short boundary lines of the object (approximate vertical lines in Figure 2.36). The problem arises for the middle line due to small gray value contrast between neighboring faces. The Hough peaks of the second stripe originate from neighboring objects at the border of the quadrangle image window.

Strategy for Applying the Vanishing-Point Compatibility

The vanishing-point compatibility is useful for slightly modifying the parameters r and ϕ of extracted image lines. A simple procedure is applied, which assumes that in a set of approximate parallel lines at least two lines are reliable and need not be adjusted. Candidates for this pair of *seed lines* are outer silhouette lines, which can be extracted robustly in case of high contrast between the gray values of object and background. Otherwise, two inner boundary lines of the silhouette could serve as seed lines as well, *e.g.* boundary lines of object faces in case of high gray value contrast due to lighting conditions or different face colors. The reliability of a line is computed on the basis of line/edge orientation-deviation in Definition 2.4. However, thus far the lines in Figure 2.36 are not restricted to the relevant line segments of the object border. Therefore, we specify for each candidate line (*e.g.* in Figure 2.36) a virtual line segment, which is of the same orientation, respectively. For the virtual segments a unique length is specified, which is assumed to be a lower bound of the lengths of all relevant boundary line segments in the image. Each virtual line segment will be moved in discrete steps along the affiliated candidate line from border to border of the polygonal window. Step by step the orientation compatibility is evaluated by applying equation (2.12). The minimum is taken as the reliability value of the line.

The most reliable two lines are selected as seed lines and their point of intersection computed, which is taken as the vanishing point. Next, the other approximate parallel lines (which are less reliable) are redefined such that they intersect at the vanishing point. Finally, the redefined lines are slightly rotated around the vanishing point in order to optimize the reliability value. In consensus with this, the weak monotony constraint must hold in the course of Hough peaks of all approximate parallel lines. Exception handling is necessary, if the two seed lines are exact parallel, because there is no finite vanishing point. In this case the unique orientation from the seed lines is adopted for the less reliable lines and a slight translation is carried out (if necessary) to optimize their reliability values. In order to take the geometric/photometric compatibility into account the seed lines and/or the redefined lines are only accepted if the line/edge orientation compatibility holds (see Assumption 2.2), otherwise they are discarded.

For example, this procedure can be applied to the set of three approximate vertical lines in Figure 2.36, represented by three non-monotonous Hough peaks in the first stripe of Figure 2.35. As a result, the two outer lines are determind as seed lines, and the inner line is slightly rotated to fulfill the vanishing-point compatibility. The next subsection introduces a further compatibility inherent in the projection of polyhedral objects, which will be applied in combination with the vanishing-point compatibility later on.

2.4.3 Pencil Compatibility of Meeting Boundary Lines

In man-made objects, the most prominent type of junction is a pencil of three lines (3-junction), respectively, e.g. a parallelepiped includes eight 3-junctions. By means of projective transformation, some parts of an opaque object boundary will be occluded, which makes certain junctions just partly visible or even invisible. For example, under general view conditions we realize in the image of a parallelepiped four 3-junctions and three 2-junctions (see Figure 2.37). That is, in four junctions all three converging lines are visible, in three junctions only two lines are visible, respectively, and one junction is completely occluded. A thorough analysis of visibility aspects of polyhedral objects was presented by Waltz for the purpose of interpreting line drawings [177, pp. 249-281]. We introduce a geometric compatibility related to 3-junctions for which three converging lines are visible, respectively.

Assumption 2.13 (Pencil compatibility) *Let a 3D pencil point be defined by the intersection of three meeting border lines of an approximate polyhedral 3D object. The projective transformation of the 3D pencil point should yield just one 2D pencil point in the image.*

For illustration, we select from the image in Figure 2.36 a subset of three boundary lines and compute the intersection points as shown in Figure 2.38. Obviously, Assumption 2.13 does not yet hold because just one common intersection point is expected instead of three. The reason is that line extraction via Hough transformation is more or less inaccurate (like any other approach to line extraction). Actually, correctness and accuracy of lines can only be evaluated with regard to the higher goal of extracting the whole object boundary. The previously introduced vanishing-point compatibility provided a first opportunity of including high level goals to line extraction, and the pencil compatibility is a second one.

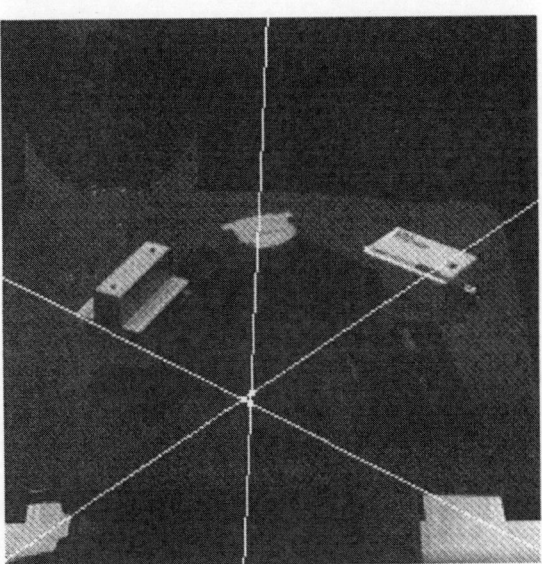

Fig. 2.38. Subset of three boundary lines taken from Figure 2.36 and three different intersection points. One unique intersection point is requested in order to fulfill the pencil compatibility.

Strategy for Applying the Pencil Compatibility

In order to make Assumption 2.13 valid, we apply a simple procedure, which only adjusts the position parameter r of image lines. The idea is to select from a 3-junction the most reliable two lines (using the procedure mentioned above), compute the intersection point, and translate the third line into this point. The approach proved to be reasonable, which is because of our frequent observation that two lines of a 3-junction are acceptable accurate and sometimes just the third line is deviating to a larger extent. For example, the most reliable two lines in Figure 2.38 are the slanted ones, and therefore the intersection point is computed and the approximate vertical line is parallel translated into this point. More sophisticated procedures are conceivable, which flexibly fine-tune the parameters of several relevant lines in combination (not treated in this work).

2.4.4 Boundary Extraction for Approximate Polyhedra

The geometric/photometric compatibility constraints and the geometric grouping criteria at the primitive, structural, and assembly level can be combined in a generic procedure for extracting the arrangement of polygons for a polyhedral object boundary. A precondition for the success of this procedure is that all relevant line segments, which are included in the arrangement of polygons, can be detected as peaks in the Hough image.

Assumption 2.14 (High gray value contrast between object faces)
All transitions between neighboring faces of a polyhedral object are characterized by high gray value contrast of at least δ_{13}.

Generic Procedure PE_3 for Extracting Arrangements of Polygons

---

Procedure PE_3

---

1. Apply the windowed OHT in a polygonal image window, detect a certain number of Hough peaks, and consider that they must be organized in stripes.
2. For each stripe of Hough peaks, examine the vanishing-point compatibility, and if it does not hold, then apply the procedure mentioned previously.
3. Compute intersection points for those pairs of image lines which are specified by pairs of Hough peaks located in different stripes.
4. Determine all groups of three intersection points in small neighborhoods. For each group examine the pencil compatibility, and if it doesn't hold, then apply the procedure mentioned previously.
5. Based on the redefined lines, determine a certain number of most salient polygons (see Definition 2.15) by applying a procedure similar to the one presented in Section 2.3.
6. Group the polygons into arrangements, compute an *assembly value* for each arrangement, and based on this, select the most relevant arrangement of polygons.

The pre-specified number of peaks to be extracted in the first step of the procedure must be high enough such that all relevant peaks are included. Another critical parameter is involved in the fifth step, *i.e.* extracting a certain number of most salient polygons. We must be careful that all relevant polygons are included which are needed for constructing the complete arrangement of polygons of the object boundary. The final step of the procedure will be implemented dependent on specific requirements and applications. For example, the grouping of polygons can be restricted to arrangements which consist of connected, non-overlapping polygons of a certain number, *e.g.* arrangements of 3 polygons for describing three visible faces of an approximate parallelepiped. The *assembly value* of an arrangement of polygons can be defined as the *mean saliency value* of all included polygons.

Experiments to the Extraction of Polygon Arrangements

With this specific implementation the generic procedure has been applied to the quadrangle image window in Figure 2.24. The extracted boundary for the object of approximate parallelepiped shape is shown in Figure 2.39. Furthermore, the procedure has been applied to more general octagons, *e.g.* a loudspeaker, whose surface consists of rectangles and trapezoids (see Figure 2.40). The generic procedure also succeeds for more complicated shapes such as the computer monitor, which has already been treated in the previous section (see Figure 2.29). The extracted boundary in Figure 2.41 demonstrates the usefulness of the pencil compatibility. There are four 3-junctions with unique pencil points, respectively (as opposed to non-unique points in Figure 2.29).

Although the procedure for boundary extraction yields impressive results for more or less complicated objects, however it may fail in simple situations. This is due to the critical assumption that all line segments of the boundary must be detected explicitly as peaks in the Hough image. For images of objects with nearly homogeneous surface color, such as the dummy box in Figure 2.39, the contrast between faces is just based on lighting conditions, which is an unreasonable basis for boundary extraction. On the other hand, for objects with surface texture or inscription spurious gray value edges may exist, which are as distinctive as certain relevant edges at the border of the object silhouette. However, all linear edge sequences produce a Hough peak, respectively. In consequence of this, perhaps a large number of Hough peak must be extracted such that all relevant boundary lines are included.

2.4.5 Geometric Reasoning for Boundary Extraction

This section presents a modified procedure for boundary extraction which applies a sophisticated strategy of geometric reasoning. It is more general in the sense that the critical Assumption 2.14, involved in the procedure presented above, is weakened. However, boundary extraction is restricted to objects of approximate parallelepiped shape and therefore the procedure is

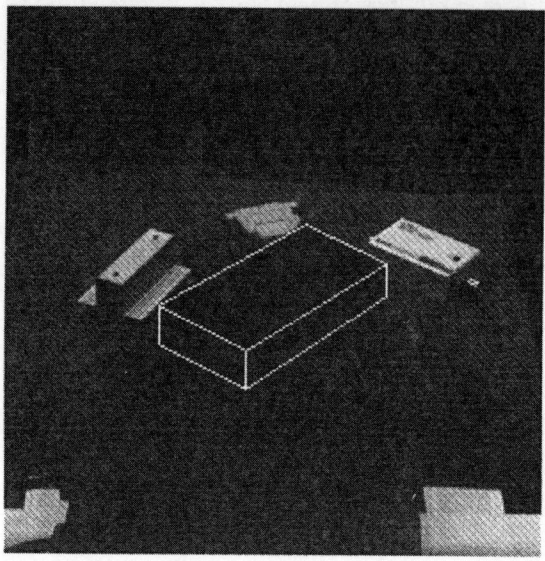

Fig. 2.39. Dummy box with approximate right-angled parallelepiped shape in a complex environment. Arrangement of polygons describing the visible boundary.

Fig. 2.40. Loudspeaker with approximate octagonal shape in a complex environment. Arrangement of polygons describing the visible boundary.

more specific concerning the object shape. The usability of the procedure is based on the following assumption.

Fig. 2.41. Computer monitor with approximate polyhedral shape including non-convexities. Arrangement of polygons describing the visible boundary.

Assumption 2.15 (Parallelepiped approximation) *The reasonable type of shape approximation for the object in a quadrangle image window is the parallelepiped.*

Generic Procedure PE_4 for Boundary Extraction of Parallelepipeds

Procedure PE_4

1. Determine a quadrangle image window which contains an object of approximate parallelepiped shape.
2. Determine just the boundary of the object silhouette which is assumed to be the most salient hexagon.
3. Propagate the silhouette lines (outer boundary lines) to the interior of the silhouette to extract the inner lines. Apply the geometric/photometric compatibility criteria and the assembly level grouping criteria to extract the most relevant arrangement of polygons.

Experiments to the Boundary Extraction of Parallelepipeds

Figure 2.42 shows the quadrangle image window containing the relevant target object, *i.e.* a transceiver box of approximate parallelepiped shape. The boundary line segments of the parallelepiped silhouette must form a hexagon (see Figure 2.43). A saliency measure is defined for hexagons, which takes into account the structural level grouping criterion of reflection-symmetry and the aspect that the hexagon must touch a large part of the quadrangle contour. This yields the boundary line segments in Figure 2.43, which are organized as three pairs of two approximate parallel line segments, respectively. Additionally, three inner line segments of the silhouette are needed to build the arrangement of polygons for the boundary of the parallelepiped. The vanishing-point compatibility is taken into account to propagate the approximate parallelism of outer lines to the interior of the silhouette. Furthermore, the pencil compatibility constrains inner lines to go through the pencil points of the silhouette boundary lines and additionally to intersect in the interior of the silhouette at just one unique point. The final arrangement of polygons must consist of just four 3-junctions and three 2-junctions. The combined use of the assembly level criteria guarantees that only two configurations of three inner lines are possible (one configuration is shown in Figure 2.44). The relevant set of three inner line segments is determined based on the best geometric/photometric compatibility. Figure 2.45 shows the final boundary line configuration for the transceiver box.

Further examples of relevant object boundaries are given below (see Figure 2.46 and Figure 2.47). They have been extracted from usual images of electrical scrap using the procedure just introduced.

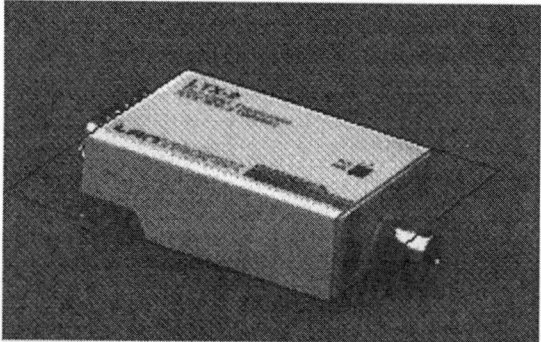

Fig. 2.42. Transceiver box with approximate right-angled parallelepiped shape. The black quadrangle surrounding the object indicates the image window for detailed processing.

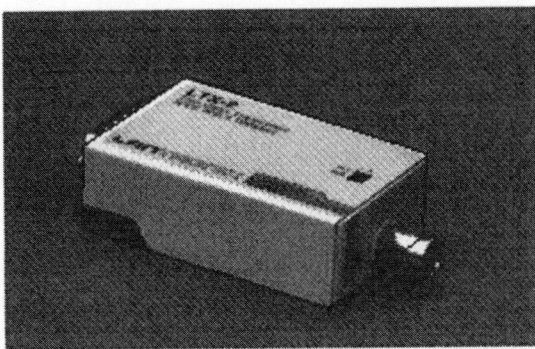

Fig. 2.43. Extracted regular hexagon, which describes the approximate silhouette of the transceiver box.

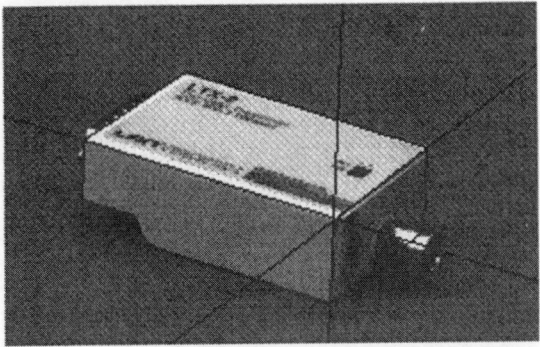

Fig. 2.44. Relevant set of three inner lines of the silhouette of the transceiver box. They have been determined by propagation from outer lines using assembly level grouping criteria and the geometric/photometric compatibility.

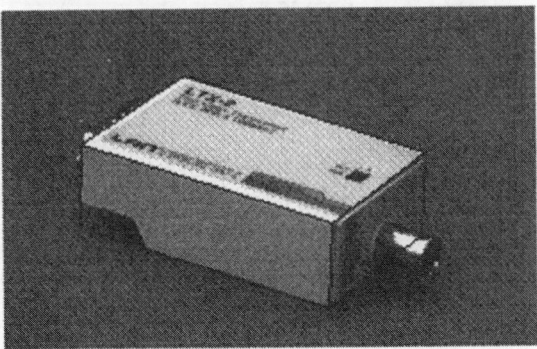

Fig. 2.45. Transceiver box with final polygon arrangement for the parallelepiped boundary description.

Fig. 2.46. Radio with approximate right-angled, parallelepiped shape and extracted arrangement of polygons of the boundary.

Fig. 2.47. Chip-carrier with approximate right-angled, parallelepiped shape and extracted arrangement of polygons of the boundary.

2.5 Visual Demonstrations for Learning Degrees of Compatibility

It is required to provide a justification for the applied compatibilities. The degrees of compatibility must be learned in the actual environment for the actual task, which is done on the basis of visual demonstration. In this section we focus on two types of geometric/photometric compatibilities, *i.e.* line/edge orientation compatibility and parallel/ramp phase compatibility. Furthermore, one type of geometric compatibility under projective transformation is treated, *i.e.* the parallelism compatibility.

2.5.1 Learning Degree of Line/Edge Orientation Compatibility

The applicability and the success of several approaches depend on accurate estimations of edge orientation. The orientations of gray value edges have been determined by applying to the image a set of four differently oriented 2D Gabor functions. Gabor parameters are the eccentricity values $\{\sigma_1, \sigma_2\}$ of the enveloping Gaussian and the center frequencies $\{u_1, u_2\}$ of the complex wave. Specific values for the Gabor parameters have influence on the estimation of edge orientations. The accuracy of edge orientation must be considered in several assumptions presented in Sections 2.2 and 2.3, *i.e.* in geometric/photometric compatibility principles and for compatibility-based structural level grouping criteria. More concretely, the accuracy of edge orientation plays a role in threshold parameters δ_1, δ_2, δ_3, δ_6, δ_7, δ_8. According to this, values for these threshold parameters must be determined on the basis of the accuracy of estimated edge orientations.

The purpose of visual demonstration is to find for the Gabor function an optimal combination of parameter values which maximizes the accuracy of orientation estimation. For explaining just the principle we make a systematic variation of just one Gabor parameter, *i.e.* the radial component of the center frequency vector ($c_r := \arctan(u_1, u_2)$) and keep the other parameters fixed. A statistical approach is used which considers patterns of different orientations. For each rotated version the edge orientations are estimated under different Gabor parametrizations. The variance of estimated edge orientations relative to the required edge orientations gives a measure for the accuracy of estimation.

Experiments to the Accuracy of Estimated Edge Orientations (Simulations)

Exemplary, we define 20 operator versions for orientation estimation (according to equations (2.6), (2.7), (2.8), (2.9), (2.10) respectively) by systematic varying the radial center frequency c_r of the involved Gabor functions. For example, Figure 2.48 shows for the Gabor functions of two operators the real part of the impulse responses, *i.e.* operator version 5 and 13.

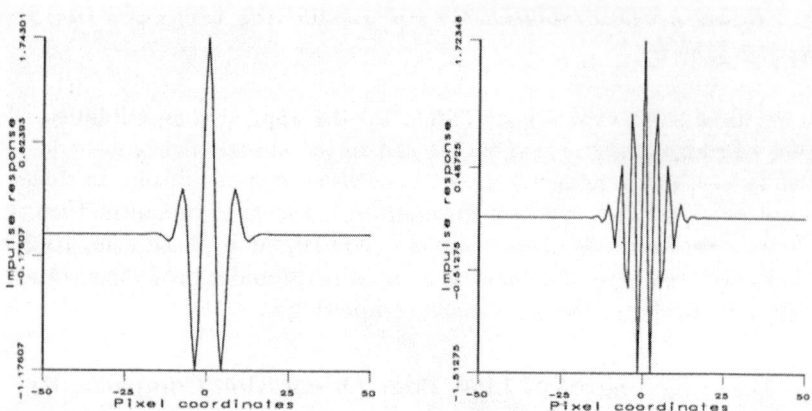

Fig. 2.48. Impulse response (real part) of two Gabor functions with different radial center frequencies.

A series of 15 simulated images of 128×128 pixels is created which consist of a black square and a gray environment, respectively. The middle point of a side of the square is the image center point which serves as turning point for the square (see white dot). The square has been rotated in discrete steps of 5° from 10° to 80°. Figure 2.49 shows a subset of four images at rotation angles 10°, 30°, 60°, 80°.

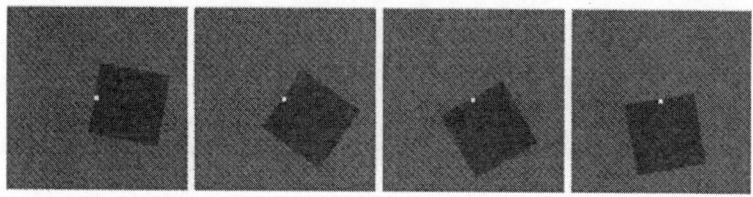

Fig. 2.49. Quadrangle in four orientations from a series of 15 orientations.

All 20 operators are applied at the center point of the 15 images in order to estimate the orientation of the edge. The accuracy of edge orientation depends on the radial center frequency. For example, Figure 2.50 shows for the black square rotated by 30° and rotated by 60° the estimations of edge orientation under varying radial center frequency. According to the two examples, the best estimation for the edge orientation is reached between frequency indices 4 and 10.

In order to validate this hypothesis we take all 15 images into account. The respective course of orientation estimations is subtracted by the relevant angle of square rotation which results in modified courses around value 0. In consequence of this, the estimation errors can be collected for all 15 images.

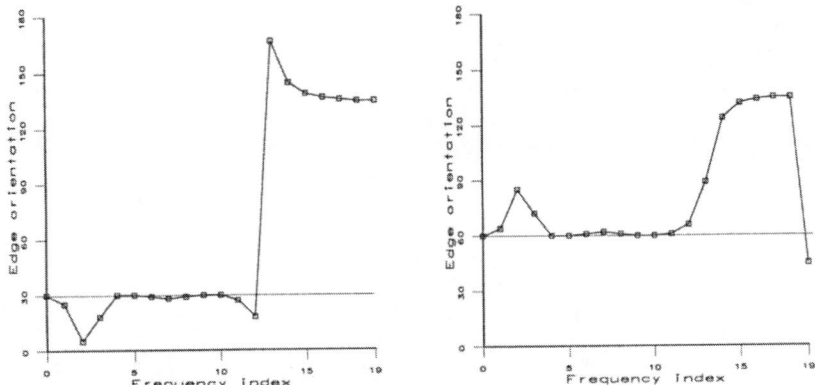

Fig. 2.50. Estimation of edge orientation at image center point under varying radial center frequency; (Left) Diagram for black square rotated by 30°; (Right) Diagram for black square rotated by 60°.

For each of the 20 operators a histogram of deviations from 0 is determined. For example, Figure 2.51 shows two histograms for operator versions 5 and 13.

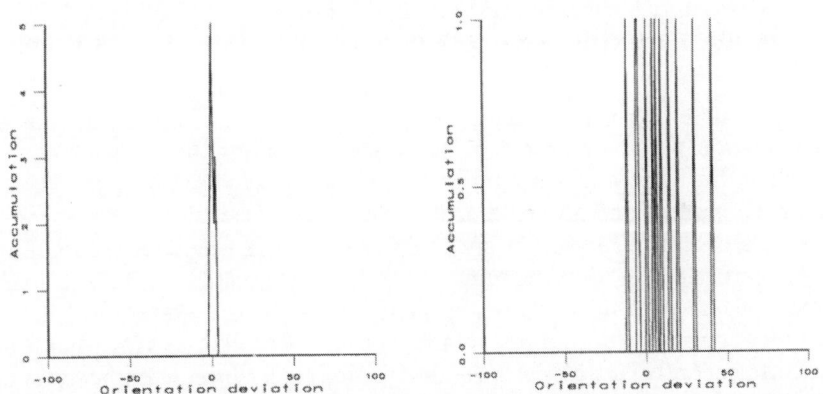

Fig. 2.51. Histograms of estimation errors for edge orientation; (Left) Diagram for operator version 5; (Right) Diagram for operator version 13.

The histogram of operator version 13 is more wide-spread than that of version 5. This means that the use of operator version 5 is favourable, because the probability for large errors in orientation estimation is low. For each of the 20 operators the variance relative to expected value 0 has been determined. Figure 2.52 shows on the left the course of variances, and on the right a section thereof (interval between frequency indices 4 and 8) in higher variance resolution. Obviously, the minimum variance of approximately 1.7 is reached at index 5, and therefore operator version 5 gives the most accurate estimations of edge orientations.

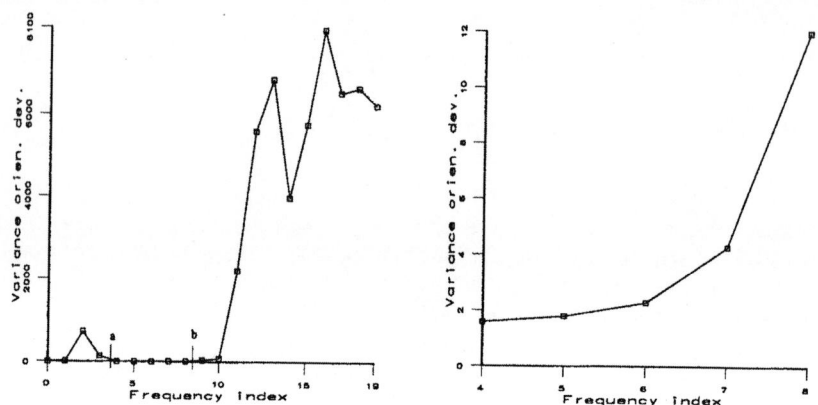

Fig. 2.52. (Left) Variance of estimation errors of edge orientation for 20 operators; (Right) Higher variance resolution of relevant section.

Experiments to the Accuracy of Estimated Edge Orientations (Real Images)

So far, those experiments have been executed for simulated image patterns. For certain parametrizations of the Gabor functions the experiments make explicit the theoretical accuracy of estimating edge orientation. However, in real applications one or more cameras are responsible for the complex process of image formation and consequently the practical accuracy must be determined. In order to come up with useful results, we have to perform realistic experiments as shown by the following figures exemplary.

All experiments from above are repeated with real images consisting of a voltage controller (taken from electrical scrap). Once again, a series of 15 images of 128 × 128 pixels is created showing the voltage controller under rotation in discrete steps of 5° from 100° to 170°. Figure 2.53 shows a subset of four images with the rotation angles 100°, 120°, 150°, 170°, and a white dot which indicates the image position for estimating edge orientation.

Figure 2.54 shows for the voltage controller rotated by 120° and rotated by 150° the estimations of edge orientation under varying radial center frequency. For the two examples the best estimation of edge orientation is reached between frequency indices 4 and 10 respectively between 0 and 7.

The estimation errors must be collected from all 15 example orientations. For example, Figure 2.55 shows two histograms of estimation errors arising

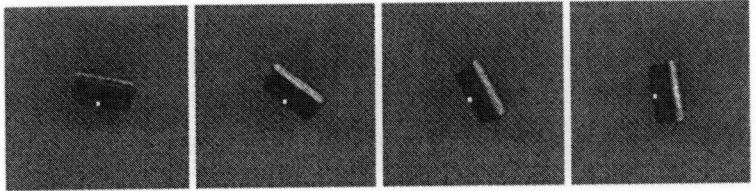

Fig. 2.53. Voltage controller in four orientations from a series of 15 orientations.

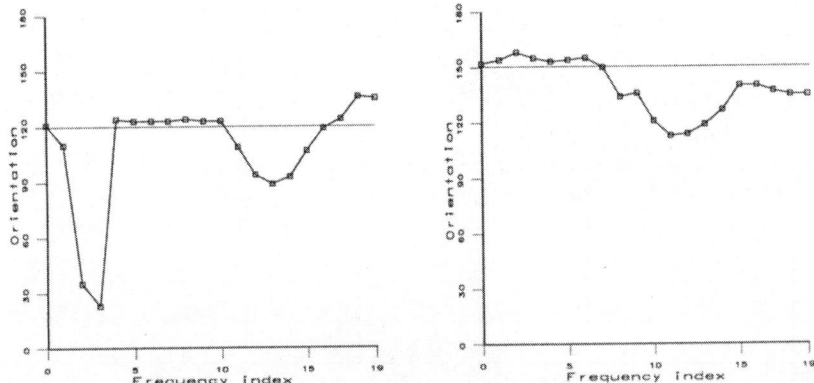

Fig. 2.54. Estimation of edge orientation at a certain point in the image of the voltage controller under varying radial center frequency; (Left) Diagram for rotation by 120°; (Right) Diagram for rotation by 150°.

from operator versions 5 and 13, respectively. From this realistic experiment we observe that the variance of estimation errors is larger than the one of the simulated situation which was depicted in Figure 2.51.

For each of the 20 operators the variance of orientation deviation has been determined. Figure 2.56 shows the course of variances on the left, and a section thereof (interval between frequency indices 4 and 6) in higher variance resolution on the right. Obviously, a minimum variance of approximately 20 is reached at indices 5 and 6, and the relevant operator versions will yield the most accurate estimations of edge orientations.

Determining Values for Threshold Parameters

The large difference between the minimum variances in the realistic case (value 20) and the simulated case (value 1.7, see Figure 2.52) motivates the necessity of executing experiments under actual imaging conditions. We can take the realistic variance value 20 to specify threshold parameters (mentioned in the beginning of this section and introduced in previous sections). For example, in our experiments it has proven useful to take for threshold δ_1 the approximate square root of the variance value, *i.e.* value 5.

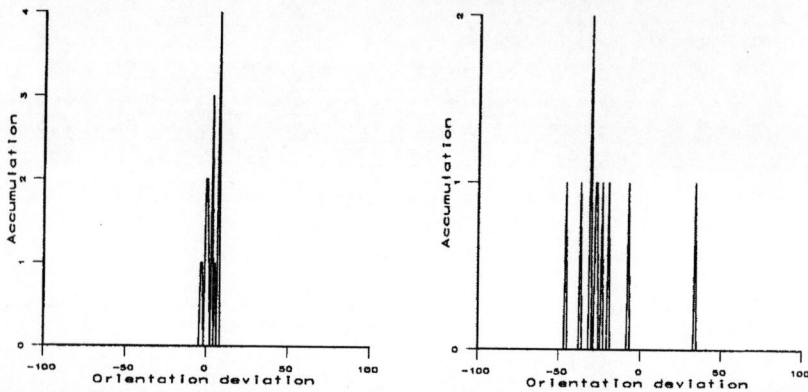

Fig. 2.55. Histograms of estimation errors for edge orientation; (Left) Diagram for operator version 5; (Right) Diagram for operator version 13.

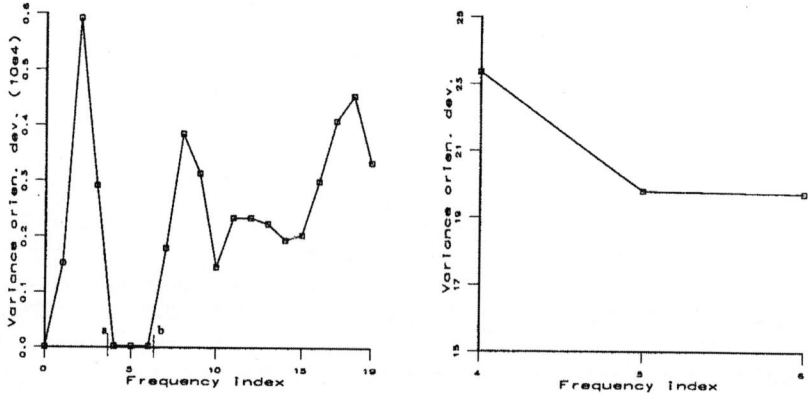

Fig. 2.56. Variance of estimation errors of edge orientation for 20 operators.

Instead of requiring that edge orientation must be computed perfectly, we made experiments in order to learn realistic estimations. These are used to determine a degree of compatibility between the orientation of a line and the orientations of edges along the line.

2.5.2 Learning Degree of Parallel/Ramp Phase Compatibility

In Section 2.3, the characteristic of the local phase has been exploited to support the grouping of those approximate parallel line segments which belong to an object silhouette. The criterion is based on a theoretical invariance, *i.e.* that the local phase computed at an edge point in the direction of the gradient angle is constant when rotating the image pattern around this edge point.

However, even for simulated images the gradient angle can not be determined exactly (see previous experiments for learning the degree of line/edge orientation compatibility). Therefore, certain variations of the local phases must be accepted.

Experiments to the Accuracy of Local Phase Computation

In the first experiment, let us first consider the series of 15 simulated images which consist of a black square touching and rotating around the image center (see Figure 2.49 for a subset of four images). The local phase computation at the image center in the direction of the estimated gradient angle yields a distribution as shown in the left diagram of Figure 2.57. The sharp peak close to value $\frac{\pi}{2}$ indicates a ramp edge at the image center.

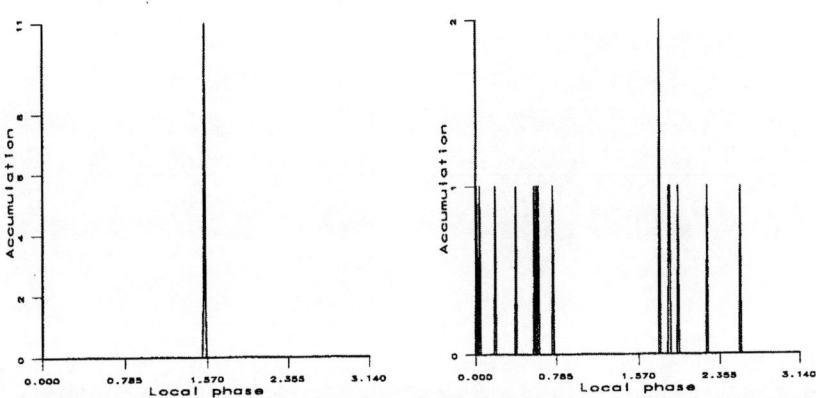

Fig. 2.57. Distribution of local phases computed at a certain point in the depiction of an object; (Left) Diagram arises for the simulated image of black square; (Right) Diagram arises for the real image of a voltage controller.

In the second experiment, we applied local phase computation to the series of real images consisting of the rotating voltage controller (see Figure 2.53 for a subset of four images). For this realistic situation a completely different distribution of local phases is obtained (see right diagram of Figure 2.57) in comparison with the simulated situation (see left diagram of Figure 2.57). The large variance off the ideal value $\frac{\pi}{2}$ arises from the sensitivity of local phase computation under selection of the image position and adjustment of the center frequency. Based on visual demonstration, we extract in the following experiments useful data for parametrizing a more robust approach of local phase computation.

Influence of Image Position and Gabor Center Frequency

We clarify the sensitive influence of image position and Gabor center frequency on the local phase estimation. This will be done for a ramp edge exemplary. At the center position of the ramp the local phase should be estimated as $\frac{\pi}{2}$ respective $-\frac{\pi}{2}$, depending on whether the ramp is upstairs or downstairs. Figure 2.58 shows on the left an image consisting of the voltage controller together with an overlay of a horizontal white line segment which crosses the boundary of the object. In the diagram on the right, the gray value structure along the virtual line segment is shown which is a ramp going downstairs (from left to right). Furthermore, the center position of the ramp edge is indicated which marks a certain point on the object boundary.

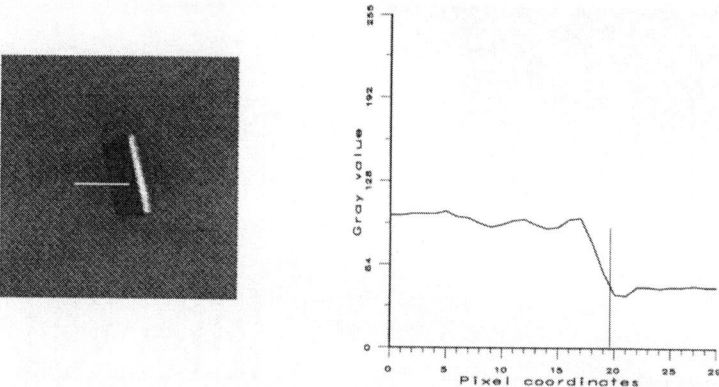

Fig. 2.58. (Left) Example image of the voltage controller with virtual line segment; (Right) Gray value structure along the virtual line segment and position of the ramp edge.

For simplifying the presentation it is convenient to project the phase computation onto the first quadrant of the unit circle, *i.e.* onto the circle bow up to length $\frac{\pi}{2}$. Along the virtual line segment in the left image of Figure 2.58 just one ramp edge is crossing. Consequently, the course of projected local phases along the virtual line segment should be an unimodal function with the maximum value $\frac{\pi}{2}$ at the center position of the ramp edge. Actually, we computed this modified local phase for each discrete point on the virtual line segment and repeated the procedure for different Gabor center frequencies. Considering the assumptions, it is reasonable to take the position of the maximum of the local phase along the virtual line segment as the location of the edge. Figure 2.59 shows four courses of local phases for four different center frequencies. There is a variation both in the position and in the value of the maximum when changing the Gabor center frequency. In the diagrams the desired position and desired local phase are indicated by a vertical and horizontal line, respectively. The higher the center frequency the less appropriate is the operator to detect a ramp edge. This becomes obvious by comparing the one distinguished global maximum produced by operator version 5 (left diagram, bold course) with the collection of local maxima produced by operator version 14 (right diagram, dotted course). However, it happens that the value of the global maximum from version 14 is nearer to the desired value $\frac{\pi}{2}$ than the value arising from operator version 5. Furthermore, the maximum values for the local phases can differ to a certain extent when applying operators with similar center frequencies, *e.g.* comparing results from operator versions 5 and 7 (left diagram, bold and dotted course).

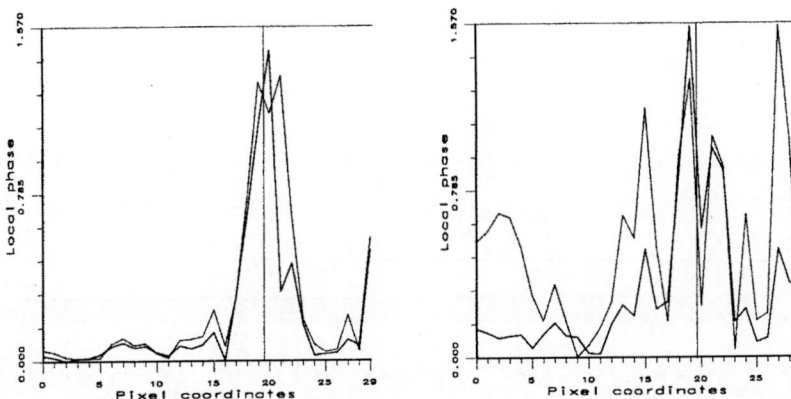

Fig. 2.59. Courses of local phases along the line segment of Figure 2.58; (Left) Courses for frequency indices 5 (bold) and 7 (dotted); (Right) Courses for frequency indices 10 (bold) and 14 (dotted).

Determining Useful Operators for Local Phase Computation

A principled treatment is required to obtain a series of useful operators. We systematical increase the Gabor center frequency and altogether apply a bank of 25 operator versions (including indices 5, 7, 10, 14 from above) to the line segment of Figure 2.58 (left). From each of the 25 courses of local phases the position and value of the global maximum will be determined. Figure 2.60 shows in the diagram on the left the course of position of the global maximum and in the diagram on the right the course of the value of the global maximum for varying center frequencies. In the left diagram the actual position of the edge is indicated by a horizontal line. Based on this, we determine a maximal band width of center frequencies for which the estimated edge position deviates from the actual one only to a certain extent. For example, if accepting a deviation of plus/minus two pixel, then it is possible to apply operator versions 0 up to 13. In the right diagram the desired local phase $\frac{\pi}{2}$ is indicated by a horizontal line. Based on this, we determine a maximal band width of center frequencies for which the estimated phase deviates from the desired one only to a certain extent. For example, if accepting a deviation of maximal 20 percent, then it is possible to apply operator versions 0 up to 23. Based on these two constraints, the series of appropriate operator versions is determined by interval intersection, *i.e.* resulting in operator versions 0 up to 13. Based on the phase estimations from these subset of relevant operators, we compute the mean value for the purpose of robustness, *e.g.* resulting in this case to 1.45.

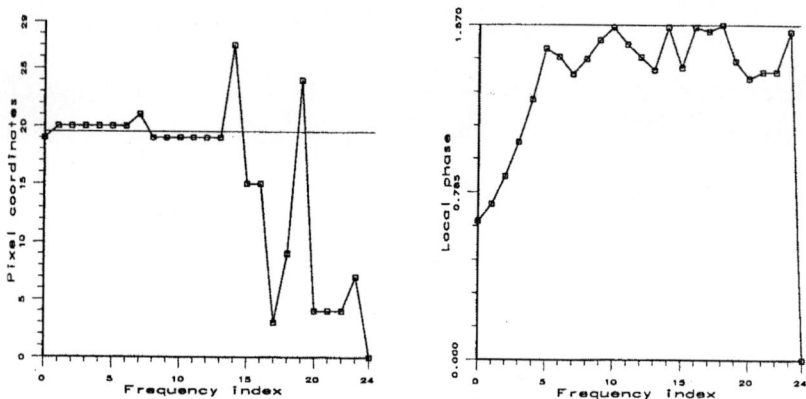

Fig. 2.60. Courses of estimated edge position (left) and local phase (right) under varying center frequency of the operator, determined along the line segment for the image in Figure 2.58.

In order to validate both the appropriateness of the operators and the mean value of phase estimation, we have to rotate the voltage controller and repeat the procedure again and again. For example, under a clock-wise rotation by 5° (relative to the previous case) we obtain appropriate operator versions 6 up to 16, and the mean value of phase estimation is 1.49. Alternatively, a counter clock-wise rotation by 5° (relative to the first case) reveals appropriate operator versions 9 up to 17, and the mean value of phase estimation is 1.45. As a result of the experimentation phase, it makes sense to determine those series of operators which are appropriate for all orientations.

Determining Values for Threshold Parameters

Based on the maximal accepted or actual estimated deviation from $\frac{\pi}{2}$, we can determine an appropriate value for threshold δ_6 (see Assumption 2.5), which represents the degree of parallel/ramp phase compatibility. Furthermore, based on the experiments we can specify the area around an edge point in which to apply local phase computation. For example, a maximal deviation of plus/minus 2 pixel has been observed, and actually this measure can be used as stripe along an image line which has been extracted by Hough transformation.

2.5.3 Learning Degree of Parallelism Compatibility

The perspective projection of parallel 3D lines yields approximate parallel lines in the image. A basic principle of structural level grouping is the search for these approximate parallel lines (see Section 2.3). The degree of deviation from exact parallelism must be learned on the basis of visual demonstration. For this purpose we mount the task-relevant objective, put the camera at a task-relevant place, and take images from a test object under varying rotation angle. An elongated, rectangular paper is used as test object with the color in clear contrast to the background. From the images of the rotating object we extract a certain pair of object boundary lines, *i.e.* the pair of approximate parallel lines which are the longest. Orientation-selective Hough transformation can be applied as basic procedure (see Section 2.2) and the relevant lines are determined by searching for the two highest peaks in the Hough image. From the two lines we take only the polar angles, compute the absolute difference and collect these measurements for the series of discrete object rotations. In the experiments, the rectangle has been rotated in discrete steps of 10° from approximately 0° to 180°.

Experiments to the Perspective Effects on Parallelism

In the first experiment, an objective with focal length $24mm$ has been used, and the distance between camera and object was about $1000mm$. Figure 2.61 (top) shows a subset of four images at the rotation angles 10°, 50°, 100°, 140°, and therein the extracted pairs of approximate parallel lines. For the whole series of discrete object rotations the respective difference between the polar angles of the lines is shown in the diagram of Figure 2.62 (left). The difference between the polar angles of the lines varies between 0° and 5°, and the maximum is reached when the elongated object is collinear with the direction of the optical axis.

For demonstrating the influence of different objectives we repeated the experiment with an objective of focal length $6mm$, and a distance between camera and object of about $300mm$. Figure 2.61 (bottom) shows a subset of four images at the rotation angles 10°, 50°, 100°, 140°, and therein the extracted pairs of approximate parallel lines. For the whole series of discrete object rotations the respective difference between the polar angles of the lines is shown in the diagram of Figure 2.62 (right). The difference between the polar angles of the lines varies between 1° and 7°. By experimentation, we quantified the increased perspective distortion for objectives of small focal length.

Taking the task-relevant experiment into account, we can specify a degree of compatibility for the geometric projection of parallel 3D lines. The

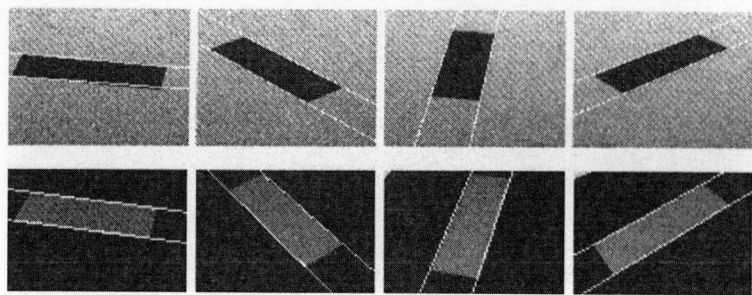

Fig. 2.61. Images taken under focal length 24mm (top row) or taken under focal length 6mm (bottom row); rectangle object in four orientations from a series of 18, and extracted pairs of approximate parallel lines.

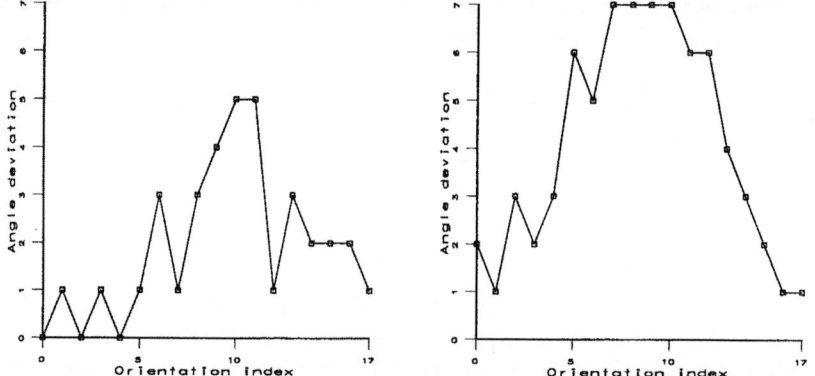

Fig. 2.62. Based on images taken under focal length 24mm (left diagram) or taken under focal length 6mm (right diagram); courses of deviations from exact parallelism for the rectangle object under rotation.

experimental data are used to supply values for the threshold parameters δ_5, δ_9, δ_{10}, δ_{11}, introduced in Section 2.3.

2.6 Summary and Discussion of the Chapter

The system for boundary extraction is organized in several generic procedures, for which the relevant definitions, assumptions, and realizations have been presented. In general, it works successful for man-made objects of approximate polyhedral shape. Interestingly, the various assumptions can be organized into three groups by considering the level of generality.

- The first group introduces general geometric/photometric compatibilities for polyhedral objects.
- The second group considers geometric compatibilities under projective transformation which hold for a subset of viewpoints.
- The third group incorporates more specific assumptions concerning object appearance and shape.

Therefore, the three groups of assumptions are stratified according to decreasing generality, which imposes a certain level of speciality on the procedures.

Validity of the Assumptions of the First Group

The assumptions of the first group should be valid for arbitrary polyhedral objects from which images are taken with usual camera objectives. This first group consists of the Assumptions 2.1, 2.2, 2.3, 2.5, 2.6, and 2.7, which are based on functions for evaluating geometric/photometric compatibilities. These compatibilities are considered between global geometric features on the one hand, *e.g.* line segment, line pencil, quadrangle, or an arbitrary polygon, and local photometric features on the other hand, *e.g.* edge orientation, corner characterization, local phase. Threshold parameters δ_1, δ_2, δ_3, δ_4, δ_6, δ_7, δ_8 are involved for specifying the necessary geometric/photometric compatibility.

These criteria can be used for accepting just the relevant line structures in order to increase the efficiency of subsequent procedures for boundary extraction. According to our experience, the parameters can be determined in a training phase prior to the actual application phase. They mainly depend on the characteristics of image processing techniques involved and of the camera objectives used. For example, we must clarify in advance the accuracy of the orientation of gray value edges and the accuracy of the localization of gray value corners, and related to the process of image formation, we are interested in the field of sharpness and the distortion effects on straight lines. Based on these measurements, we compute the actual degrees of deviation from exact invariance and conclude about acceptance of compatibilities. If a compatibility exists then the relevant threshold parameters are specified. In the case of rejection, one must consider more appropriate image processing techniques and/or other camera objectives. According to this, the role of experimentation is to test the appropriateness of certain image processing techniques in combination with certain camera objectives for extracting certain image structures.

Validity of the Assumptions of the Second Group

The assumptions of the second group are supposed to be valid for arbitrary polyhedral objects but there is a restriction concerning the acceptable view conditions. This second group consists of the Assumptions 2.4, 2.8, 2.9, 2.10, 2.12, 2.13. They impose constraints on the projective transformation of geometric features of 3D object shapes. To consider the regularity aspect of man-made objects a set of collated regularity features is used, such as parallel lines, right-angled lines, reflection-symmetric polylines, or translation-symmetric polylines. The object shapes are detected in the image as salient polygons or arrangements of polygons. Several saliency measures have been defined on the basis of geometric/photometric compatible features and the collated regularity features (just mentioned).

It is essential that also the regularity features are compatible under projective transformation. Regularity and compatibility depend on each other, *e.g.* the vanishing-point compatibility affects the necessary degree of deviation from parallelism. The degree of deviation from exact invariance depends on the spectrum of permissible camera positions relative to the scene objects. Threshold parameters δ_5, δ_9, δ_{10}, δ_{11} are involved in the assumptions for describing the permissible degrees of deviation from exact invariance. For example, if we would like to locate the right-angled silhouette of a flat object (*e.g.* an electronic board) then the camera must be oriented approximately perpendicular to this object, and this will be considered in the parameter δ_{11} (see Assumption 2.10).

Validity of the Assumptions of the Third Group

The basic assumption that the scene consists of approximate polyhedral objects usually is too general for providing one and only one generic procedure for boundary extraction. Therefore, a third group of constraints is introduced consisting of the Assumptions 2.11, 2.14, 2.15. They impose constraints on the gray value appearance and the shape of the depicted objects. We must examine whether an extracted polygon is an approximate representation of the object silhouette, or examine whether the transistions between object faces have high gray value contrast, or examine whether the shape of an object in a quadrangle image window is an approximate parallelepiped. Threshold parameters δ_{12} and δ_{13} are involved for quantifying these constraints. Although the assumptions of this third group are more specific than those of the other two groups, they are somewhat general.

Gaussian Assumption Related to Feature Compatibilities

In numerous experiments we observed that the distribution of deviations from invariance is Gaussian-like, more or less. That was the motivation for comparing the concept of compatibility with the concept of invariance via the Gaussian, *i.e.* invariance is a special case of compatibility with sigma equal 0. More generally, the deviations from the theoretical invariant can be approximated by computing the covariance matrix and thus assuming a multi-dimensional Gaussian which maybe is not rotation-symmetric and not in normal form. The sophisticated approximation would be more appropriate for deciding whether certain procedures are applicable, and determine appropriate parameters from the covariance matrix instead of scalar sigma.

Summary of the Approach of Boundary Localization

Altogether, our approach succeeds in locating and extracting the boundary line configurations for approximate polyhedral objects in cluttered scenes. Following Occam's minimalistic philosophy, the system makes use of fundamental principles underlying the process of image formation, and makes use of general regularity constraints of man-made objects. Based on this, the role of specific object models is reduced. This aspect is useful in many realistic applications, for which it is costly or even impossible to acquire specific object models. For example, in the application area of robotic manipulation of electrical scrap (or car scrap, etc.), it is inconceivable and anyway unnecessary to explicitly model all possible objects in detail. For robotic manipulation of the objects approximate polyhedral descriptions are sufficient, which can be extracted on the basis of general assumptions. The novelty of our methodology is that we maximally apply general principles and minimally use object-specific knowledge for extracting the necessary information from the image to solve a certain task.

Future work should discover more compatibilities between geometry and photometry of image formation, and more compatible regularity features under projective transformation.[9] The combination of compatibility and regularity must be treated thoroughly, *e.g.* solving the problem of combined constraint satisfaction. An extension of the methodology beyond man-made objects, *e.g.* natural objects such as faces, is desirable.

The next chapter presents a generic approach for learning operators for object recognition which is based on constructing feature manifolds. Actually, a manifold approximates the collection of relevant views of an object, and therefore represents a kind of compatibility between object views. However, the representation of the compatibilities between various object views is more difficult compared to the compatibilities treated in this chapter.

---

[9] In Subsection 4.3.5 we will introduce another type of compatibility under projective transformations which will prove useful for treating the problem of stereo matching.

3. Manifolds for Object and Situation Recognition

This chapter presents a generic approach for learning operators for object recognition or situation scoring, which is based on constructing feature manifolds.

3.1 Introduction to the Chapter

The introductory section of this chapter describes in a general context the central role of learning for object recognition, then presents a detailed review of relevant literature, and finally gives an outline of the following sections.[1]

3.1.1 General Context of the Chapter

Famous physiologists (*e.g.* Hermann von Helmholtz) insisted on the central role of learning in visual processes [46]. However, only a few journals dedicated issues to this aspect (*e.g.* [5]), only a few workshops focused on learning in Computer Vision (*e.g.* [19]), and finally, only a few doctoral dissertations treated learning as the central process of artificial vision approaches (*e.g.* [24]). We strongly believe that the paradigm of Robot Vision must completely be arranged around learning processes at all levels of feature extraction and object recognition. The inductive theory of vision proposed in [65] is in consensus with our believe, *i.e.* the authors postulate that vision processes obtain all the basic representations via inductive learning processes. Contrary to the "school of D. Marr", machine vision is only successful by using both the input signal and, importantly, using also learned information. The authors continue with the following two statements which we should emphasize.

> *"The hope is that the inductive processes embody the universal and efficient means for extracting and encoding the relevant information from the environment."*

---

[1] The learning of signal transformations is a fundamental characteristic of Robot Vision, see Section 1.2.

J. Pauli: Learning-Based Robot Vision, LNCS 2048, pp. 101-169, 2001.
© Springer-Verlag Berlin Heidelberg 2001

> *"The evolution of intelligence could be seen, not as ad hoc, but as a result of interactions of such a learning mechanism with the environment."*

In consensus with this, a main purpose of this work is to declare that all vision procedures, *e.g.* for signal transformation, feature extraction, or object recognition, must be based on information which has been adapted or learned in the actual environment (see Section 1.4).

In Section 2.5 it was shown exemplary how to determine useful (intervals of) parameter values in order to apply procedures for boundary line extraction appropriately. Furthermore, for various types of compatibilities which do hold under real image formation, the degrees of actual deviations from invariants have been determined. All those is based on visual demonstrations in the task-relevant environment and the necessary learning process simply consists of rules for closing intervals or for approximating distributions. In this chapter the (neural) learning plays a more fundamental role. Operators must be learned for the actual environment constructively, because the compatibilities presented in the previous Chapter 2 are not sufficient to solve certain tasks. For example, we would like to recognize certain objects in the scene in order to manipulate them specifically with the robot manipulator. The learning procedure for the operators is based on 2D appearance patterns of the relevant objects or response patterns resulting from specific filter operations. The main interest is to represent or approximate the pattern manifold such that an optimal compromise between *efficiency, invariance and discriminability* of object recognition is achieved.

3.1.2 Approach for Object and Situation Recognition

This work focuses on a holistic approach for object and situation recognition which treats appearance patterns, filtered patterns, or histogram patterns (vectors) and leaves local geometric features implicit. The recognition is based on functions which are not known *a priori* and therefore have to be learned in the task-relevant environment. It is essential to keep these functions as simple as possible, as the complexity is correlated to the time needed for object or situation recognition. Three aspects are considered in combination in order to meet this requirement.

- Appearance patterns should be restricted to relevant windows, *e.g.* by taking the silhouette boundary of objects into account (see Section 2.3) in order to suppress the neighborhood or background of a relevant object.
- Only those appearance variations should be taken into account which indispensable occur in the process of task-solving, *e.g.* change of lighting condition, change of relation between object and camera, *etc.*
- For the task-relevant variety of appearance patterns we are looking for types of representation such that the manifold dimension decreases, *e.g.*

gray value normalization, band-pass Gabor filtering, log-polar transformation.

Coarse-to-Fine Strategy of Learning

Techniques for simplifying the pattern manifold are applied as a pre-processing step of the learning procedure. For learning the recognition function a *coarse-to-fine strategy* is favoured. The *coarse* part treats global and the *fine* part treats local aspects in the manifold of patterns.

First, a small set of distinguished appearance patterns (*seed patterns*) are taken under varying imaging conditions, represented appropriately, and approximated by a learning procedure. Usually, these patterns are globally distributed in the manifold, and they serve as a *canonical frame (CF)* for recognition. The algebraic representation is by an implicit function whose value is (approximately) 0 and in this sense the internal parameters of the function describe a *global compatibility between the patterns*. Computationally, the implicit function will be learned and represented by a *network of Gaussian basis functions (GBF network)* or alternatively by *principal component analysis (PCA)*.

Second, the global representation is refined by taking appearance patterns from counter situations into account, which actually leads to local specializations of the general representation. The implicit function of object recognition is modified such that compatibility between views is conserved but with the additional characteristic of discriminating more reliably between various objects. Furthermore, the recognition function can be adjusted more carefully in critical regions of the pattern manifold by taking and using additional images of the target object. For the local modification of the global manifold representation we apply once again Gaussian basis functions or principal component analysis.

In summary, our approach for object and situation recognition uses recognition functions which are acquired and represented by mixtures of GBFs and PCAs. Based on the brief characterization of the methodology we will review relevant contributions in the literature.

3.1.3 Detailed Review of Relevant Literature

A work of Kulkarni *et al.* reviews classical and recent results in statistical pattern classification [92]. Among these are nearest neighbor classifiers, the closely related kernel classifiers, classification trees, and various types of neural networks. Furthermore, the *Vapnik-Chervonenkis theory of learning* is treated. Although the work gives an excellent survey it can not be complete, *e.g.* principal component analysis is not mentioned, dynamically growing neural networks are missing.

Principal Component Analysis (PCA)

Principal component analysis (PCA) is a classical statistical method which determines from a random vector population the system of orthogonal vectors, *i.e.* so-called *principal components*, such that the projection of the random vectors onto the first component yields largest variance, and the projection onto the second component yields second largest variance, and so on [60, pp. 399-440]. The principal components and the variances are obtained by computing eigenvectors and eigenvalues of the covariance matrix constructed from the random vector population. For the purpose of classification, the eigenvectors with the largest eigenvalues are the most important one, and class approximation takes place by simply omitting the eigenvectors with the lowest eigenvalues. The representation of a vector by (a subset of most important) eigenvectors, *i.e.* the eigenspace, is called the *Karhunen-Loéve expansion (KLE)* which is a linear mapping.

Turk and Pentland applied the PCA approach to face recognition and also considered problems with different head sizes, different head backgrounds and localization/tracking of a head [168]. Murase and Nayar used PCA for the recognition of objects which are rotated arbitrary under different illumination conditions [112]. The most serious problem with PCA is the daring assumption of a multi-dimensional Gaussian distribution of the random vector population, which is not true in many realistic applications. Consequently, many approaches of *nonlinear dimension reduction* have been developed in which the input data are clustered and local PCA is executed for each cluster, respectively. A piece-wise linear approximation is obtained which is global nonlinear [89, 163]. In a work of Prakash and Murty the clustering step is more closely combined with PCA which is done iteratively on the basis of classification errors [133]. A work of Bruske and Sommer performs clustering within small catchment areas such that the cluster centers can serve as reasonable representatives of the input data [30]. Based on this, local PCA is done between topologically neighbored center vectors instead of the larger sets of input vectors.

Radial Basis Function Networks (RBF Networks)

The construction of topology preserving maps (functions) from input spaces to output spaces is a central issue in neural network learning [106]. It was proven by Hornik *et al.* that multilayer feedforward networks with arbitrary squashing function, *e.g.* MLP (multilayer perceptron) networks, can approximate any Borel-measurable function to any desired degree of accuracy, provided sufficiently many hidden units are available [81]. Similar results have been reported for so-called *regularization networks, e.g. RBF (radial basis function) networks* and the more general *HBF (hyper basis function) networks* [63], which are theoretically grounded in the regularization theory. Besides the theoretical basis the main advantage of regularization networks is the transparency of what is going on, *i.e.* the meaning of nodes and links.

A regularization network interpolates a transformation of m-dimensional vectors into p-dimensional vectors by a linear combination of n nonlinear basis functions. Each basis function, represented in a hidden node, operates as a localized receptive field and therefore responds most strongly for input vectors localized in the neighborhood of the center of the field.

For example, the spherical Gaussian is the most popular basis function which transforms the distance between an input vector and the center vector into a real value of the unit interval. Classically, the centers and the extent of the receptive fields are learned with unsupervised methods, while the factors for combining the basis functions are learned in a supervised manner. Hyper basis functions are prefered for more complex receptive fields, *e.g.* hyper-ellipsoidal Gaussians for computing the *Mahalanobis distance* to the cluster center (based on local PCA). We will introduce the unique term *GBF networks* for specific regularization networks which consist of *hyper-spherical* or hyper-ellipsoidal Gaussians. It is the transparency why specifically the GBF networks are used in dynamic network architectures, *i.e.* the hidden nodes are constructed dynamically in a supervised or unsupervised learning methodology [59, 29]. Transparency has also been a driving force for the development of *parallel consensual neural networks* [18] and *mixture-of-experts networks* [174].

Support Vector Networks (SV Networks)

Recently, the so-called *support vector networks* became popular whose construction is grounded on the principle of *minimum description length* [139], *i.e.* the classifier is represented by a minimum set of important vectors. The methodology of *support vector machines* consists of a generic learning approach for solving nonlinear classification or regression problems. The pioneering work was done by Vapnik [171], an introductory tutorial is from Burges [31], and an exemplary application to object recognition has been reported in a work of Pontil and Verri [132]. The conceptual idea of constructing the classifier is to transform the input space nonlinear into a feature space, and then determine there a linear decision boundary. The optimal hyperplane is based only on a subset of feature vectors, the so-called *support vectors (SV)*, which belong to the common margin of two classes, respectively. The relevant vectors and the parameters of the hyperplane can be determined by solving a quadratic optimization problem. As a result, the decision function is a linear combination of so-called *kernel functions* for which *Mercer's condition* must be satisfied.

An example of a kernel function is the spherical Gaussian. Accordingly, the formula implemented by a support vector machine with Gaussian kernels is identical to the formula implemented by an RBF network [160]. The distinction between an RBF network and a support vector machine with Gaussian kernels is by the technique of constructing the unknowns, *i.e.* the center vectors, the extensions and the combination factors of the Gaussians. In a work of Schölkopf *et al.* the classical approach of training an RBF network (unsupervised learning of Gaussian centers, followed by supervised learning of combination factors) is compared with the support vector principle of determining the unknowns, and as a result, the SV approach proved to be superior [151].[2]

Bias/Variance Dilemma in SV Networks

Support vector machines enable the treatment of the *bias/variance dilemma* [62] by considering constraints in the construction of the support vectors. In order to understand the principle we regard a class as a set of local transformations which have no influence on the class membership. The PhD thesis of Schölkopf presents two approaches within the support vector paradigm which consider and make use of this aspect [152, pp. 99-123]. In the first approach, the support vector machine is trained to extract the set of support vectors, then further virtual examples are generated in a localized region around the support vectors, and finally, a new support vector machine is learned from the virtual data. The second approach is similar to the first one, however, the learning procedure is not repeated with virtual data, but a regularization term is considered which is based on a known one-parameter group of local transformations, *e.g.* the tangent vectors at the set of support vectors.[3] In addition to these two approaches, which consider local transformations, a third approach is conceivable for certain applications. It is characterized by a global transformation of the input data into a new coordinate system in which class membership can be determined easily. For example, by taking the *Lie group theory* into account it is possible to construct kernels of integral transforms which have fine invariance characteristics [154]. The role of integral transforms for eliciting invariances and representing manifolds can be exemplified by the Fourier transform, *e.g.* the spectrum of a Fourier-transformed image does not change when translating a pattern in the original image.

The three approaches of enhancing the learning procedure are summarized in Figure 3.1 (adopted from the PhD thesis of Schölkopf [152, p. 101]). For simplicity a two-dimensional space is assumed, in which the dashed line depicts the desired decision boundary, and the black and gray disks represent elements from two different classes. The first approach (on the left) considers virtual examples around the actual examples, the second approach (in

---

[2] However, the comparison seems to be unfair, because one can alternatively switch to another RBF learning approach which determines not only the Gaussian combination factors but also the Gaussian center vectors by error backpropagation ([21, pp. 164-193], [122]). Furthermore and at least equal important, the comparison considers only spherical Gaussians. However, a network of hyper-ellipsoidal Gaussians, *i.e.* a GBF network with each Gaussian determined by local PCA, usually does a better job in classification tasks [128, 69].

[3] A work of Burges treats the geometry of local transformations and the incorporation of known local invariances into a support vector machine from a theoretical point of view [32].

the middle) uses tangent values around the examples, and the global transformation approach (on the right) is visualized by large circles through the examples which indicate that possibly only the radius is of interest for classification.

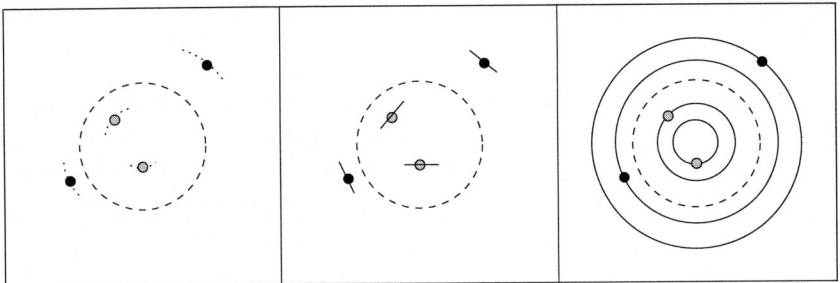

Fig. 3.1. (Left) Enhanced learning by virtual examples; (Middle) Tangent vectors; (Right) Modified representation.

Classical Invariance Concepts for Recognition Tasks

Although the concept of invariance plays a significant role in recognition, we only give a short survey. The reason is that the known approaches are only of limited use for our purpose. Mundy and Zisserman reviewed geometric invariants under transformation groups and summarized methods for constructing them [111]. However, methods such as the *elimination of transformation parameters* are not adequate for hard recognition tasks, because we do not know the transformation formula and can not assume group characteristics. Alternatively, the review given by Wechsler is more extensive, as it includes geometric, statistical, and algebraic invariances [175, pp. 95-160]. Related to appearance-based object recognition the aspect of statistical invariance is treated by principal component analysis.

Evaluating Learning Procedures and Learned Constructs

The usefulness and performance of all these learning procedures should be assessed on the basis of objective criteria in order to choose the one which is most appropriate for the actual task. However, a formal proof that a certain learning algorithm is superior compared to other ones can not be given (usually). The goal of theoretical work on learnability is to provide answers to the following questions.

> What problems can be learned, how much training data is required, how much classification errors will occur, what are the computational costs of applying the learning algorithm ?

The most popular formalization of these questions is from Vapnik and Chervonenkis who introduced the so-called *PAC-learnability* with PAC an acronym of "probably approximately correct" [170]. A function is called PAC-learnable if and only if a learning algorithm can be formulated which produces with a certain expenditure of work a probably approximately correct function. Based on pre-specified levels for probability and correctness of a function approximation, we are interested in discovering the most simple and efficient applicable representation (Occam's Razor). For example, authors of the neural network community derive lower and upper bounds on the sample size versus net size needed such that a function approximation of a certain quality can be expected [15, 6]. On account of applying Occam's Razor to GBF networks, it is desirable to discover the minimum number of basis functions in order to reach a critical quality for the function approximation.

3.1.4 Outline of the Sections in the Chapter

Section 3.2 describes the learning of functions for object recognition as the construction of pattern manifolds. Two approximation schemes are compared, *i.e.* networks of Gaussian basis functions and principal component analysis. In Section 3.3 GBF networks are applied for globally representing a manifold of object appearances under varying viewing conditions or a manifold of varying grasping situations. Based on visual demonstration, the number and the extent of the Gaussians are modified appropriately in order to obtain recognition functions with a certain quality. Section 3.4 presents alternative representations of appearance manifolds. These include a fine-tune of the Gaussians by considering topologically neighbored patterns, or alternatively a global PCA instead of a GBF network. Furthermore, the global manifold representation will be refined by taking appearance patterns from counter situations into account using GBFs once again. Section 3.5 discusses the approaches of the preceding sections.

3.2 Learning Pattern Manifolds with GBFs and PCA

This section describes the learning of functions for object recognition as the construction of pattern manifolds. Robustness of recognition is formulated in the PAC terminology. Two approximation schemes are compared, *i.e.* networks of Gaussian basis functions and principal component analysis.

3.2.1 Compatibility and Discriminability for Recognition

Usually, invariants are constructed for groups of transformations. The more general the transformations the more difficult to extract invariants of the whole transformation group. Even if it is possible to extract invariants for a general transformation group, they are of little use in practice due to their generality. For an object under different view conditions the manifold of appearance patterns is complex in nature, *i.e.* general transformations between the patterns. Instead of determining invariants for a geometric transformation group, we are interested in *compatibilities among object appearances*. The compatibilities should be as specific as necessary for discriminating a target object from other objects. The actual task and the relevant environment are fundamental for constructing an appropriate recognition function. First, as a preprocessing step the pattern manifold should be simplified by image normalization, image filtering, and representation changing. This strategy is applied several times throughout the sections in this and the next chapter. Second, only the relevant subset of transformations should be considered which must be learned on the basis of visual demonstration. Formally, the learning step can be treated as parameter estimation for implicit functions.

Invariance Involved in Implicit Functions

Let f^{im} be an implicit function with parameter vector B and input-output vector Z, such that

$$f^{im}(B, Z) = 0 \qquad (3.1)$$

The parameter vector B characterizes the specific version of the function f^{im} which is of a certain type. The vector Z is called the input-output vector in order to express that input and output of an explicitly defined function are collected for the implicitly defined function. Input-output vector Z is taken as variable and we are interested in the manifold of all realizations for which equation (3.1) holds. In order to introduce the term *compatibility* into the context of *manifolds*, we may say, that all realizations in the manifold are compatible to each other.

For example, for all points of a two-dimensional ellipse (in normal form) the equation (3.1) holds if

$$f^{im}(B, Z) := \frac{x_1^2}{b_1^2} + \frac{x_2^2}{b_2^2} - 1, \qquad (3.2)$$

with parameter vector $B := (b_1, b_2)^T$ containing specifically the two half-lengths b_1 and b_2 of the ellipse axes in normal form, and vector $Z := (x_1, x_2)^T$ containing the 2D coordinates. In terms of the *Lie group theory of invariance* [125], the manifold of realizations of Z is the orbit of a differential operator which is generating the ellipse. Function f^{im} is constant for all points of the elliptical orbit, and therefore the half-lengths of the ellipse axes are *invariant features* of the responsible generator.

Implicit Functions for Object or Situation Recognition

For the purpose of object or situation recognition the function f^{im} plays the role of a recognition function and therefore is much more complicated than in the previous example. The variation of the real world environment, *e.g.* object translation and/or rotation, background and/or illumination change, causes transformations of the appearance pattern, whose task-relevant orbit must be approximated. In the next two sections we present two approaches for approximating the recognition function, *i.e.* networks of Gaussian basis functions (GBF approach) and principal component analysis (PCA approach). In the GBF approach the parameter vector B comprises positions, extents, and combination factors of the Gaussians. In the PCA approach the parameter vector B comprises the eigenvectors and eigenvalues. The input-output vector Z is separated in the input component X and the output component Y, with the input part representing appearance patterns, filtered patterns, or histogram patterns of objects or situations, and the output part representing class labels or scoring values.

The Role of Inequations in Implicit Recognition Functions

In order to apply either the GBF approach or the PCA approach, the implicit function f^{im} must be learned in advance such that equation (3.1) holds more or less for patterns of the target object and clearly not holds for patterns of counter situations. Solely small deviations from the ideal orbit are accepted for target patterns and large deviations are expected for counter patterns. The degree of deviation can be controlled by a parameter ψ.

$$| f^{im}(B, Z) | \leq \psi \tag{3.3}$$

The function f^{im} can be squared and transformed by an exponential function in order to obtain a value in the unit interval.

$$f^{Gi}(B, Z) := exp\left(-f^{im}(B, Z)^2\right) \tag{3.4}$$

If function f^{Gi} yields value 0, then vector Z is infinite far away from the orbit, else if function f^{Gi} yields value 1, then vector Z belongs to the orbit. Equation (3.1) can be replaced equivalently by

$$f^{Gi}(B, Z) = 1 \tag{3.5}$$

For reasons of consistency, we also use the exponential function to transform parameter ψ into ζ. Parameter ψ was a threshold for distances, but parameter ζ is a threshold for proximities.

$$\zeta := exp\left(-\psi^2\right) \tag{3.6}$$

With this transformations, we can replace equation (3.3) equivalently by

$$f^{Gi}(B, Z) \geq \zeta \tag{3.7}$$

Based on these definitions, we extend the usage of the term *compatibility*, and embed it in the concept of manifolds.

Assumption 3.1 (Manifold compatibility) *Let function f^{Gi} and parameter vector B be defined as above, and let threshold ζ be the minimal proximity to the ideal orbit of f^{Gi}. Then all realizations of Z for which equation (3.7) holds are compatible to each other.*

In the application of object or situation recognition, we must embed the compatibility criterion into the requirement for a robust operator. The *robustness of recognition* is defined by incorporating an *invariance* criterion and a *discriminability* criterion. The invariance criterion strives for an operator which responds nearly equal for any pattern of the target object, *i.e.* compatibility criterion with threshold ζ near to 1. The discriminability criterion aims at an operator, which clearly discriminates between the target object and any other object or situation. Regions of the appearance space, which represent views of objects other than the target object or any background area, should be given low confidence values. The degree of robustness of an operator for object or situation recognition can be specified more formally in the PAC terminology.

Definition 3.1 (PAC-recognition) *Let parameter ζ define a threshold for a certain proximity to the ideal orbit. A function f^{Gi} for object or situation recognition is said to be PAC-learned subject to the parameters P^r and ζ if, with a probability of at least P^r, the function value of a target pattern surpasses and the function value of a counter pattern falls below threshold ζ.*

In several subsections throughout this chapter, we will present refinements and applications of this generic definition of PAC-recognition. Related to the problem of learning operators for object or situation recognition, a main purpose of visual demonstration is to analyze the conflict between invariance and discriminability and find an acceptable compromise.

3.2.2 Regularization Principles and GBF Networks

An approach for function approximation is needed which has to be grounded on sample data of the input–output relation. The function approximation should fit the sample data to meet *closeness constraints* and should generalize over the sample data to meet *smoothness constraints*. Neglecting the aspect of generalizing leads to *overfitted functions*, otherwise, neglecting the fitting aspect leads to *overgeneralized functions*. Hence, both aspects have to be combined to obtain a qualified function approximation. The *regularization approach* incorporates both constraints and determines such a function by minimizing a functional [130].

Theoretical Background of GBF Networks

Let $\Omega_{GBF} := \{(X_j, y_j) | (X_j, y_j) \in (\mathcal{R}^m \times \mathcal{R}); j = 1, \cdots, J\}$ be the sample data representing the input–output relation of a function f that we want to approximate. The symbol $\mathcal{R}$ designates the set of real numbers. The functional F in equation (3.8) consists of a *closeness term* and a *smoothness term*, which are combined by a factor μ expressing relative importance of each.

$$F(f) := \left(\sum_{j=1}^{J} (y_j - f(X_j))^2 \right) + \mu \cdot \| F^{sm}(f) \|^2 \qquad (3.8)$$

The first term computes the sum of squared distances between the desired and the actual outcome of the function. The second term incorporates a differential operator F^{sm} for representing the smoothness of the function.

Under some pragmatic conditions (see again [130]) the solution of the regularization functional is given by equation (3.9).

$$f(X) := \sum_{j=1}^{J} w_j \cdot f_j^{Gs}(X) \qquad (3.9)$$

The basis functions f_j^{Gs} are *Gaussians* with $j \in \{1, \cdots, J\}$, specified for a limited range of definition, and having X_j as the centers. Based on the non-shifted Gaussian basis function f^{Gs}, we obtain the J versions f_j^{Gs} by shifting the center of definition through the input space to the places $X_1, \cdots, X_J$. The solution of the regularization problem is a linear combination of Gaussian basis functions (GBF) (see equation (3.9)).

Constructing the Set of GBFs

The number of GBFs must not be equal to the number of samples in Ω_{GBF}. It is of interest to discover the minimum number of GBFs which are needed to reach a critical quality for the function approximation. Instead of using the vectors $X_1, \cdots, X_J$ for defining GBFs, we cluster them into I sets (with $I \leq J$) striving simultaneous for minimizing the variances within and maximizing the distances between the sets. A procedure similar to the error-based ISODATA clustering algorithm can be used which results in I sets (see Subsection 2.3.1). From each set a mean vector $X_i^c, i \in \{1, \cdots, I\}$, is selected (or computed) which specifies the center of the definition range of a GBF in normal form.

$$f_i^{Gs}(X) := exp\left(-\frac{1}{\tau} \cdot \frac{\|X - X_i^c\|^2}{\sigma_i^2} \right) \qquad (3.10)$$

The function f_i^{Gs} computes a similarity value between the vector X_i^c and a new vector X. The similarity is affected by the pre-specified parameters σ_i and τ, whose multiplicative combinations determine the *extent* of the GBF. Parameter σ_i is defined by the variance of elements in the relevant cluster, averaged over all components of the m-dimensional input vectors. It is intuitive clear that the ranges of definition of the functions G_i must overlap to a certain degree in order to approximate the recognition function appropriately. This overlap between the GBFs can be controlled by the factor τ.

Alternative to spherical Gaussians, a GBF can be generalized by taking the covariance matrix $\mathcal{C}_i$ into account which describes the distribution of cluster elements around the center.

$$f_i^{Gs}(X) := exp\left(-\frac{1}{\tau} \cdot (X - X_i^c)^T \cdot \mathcal{C}_i^{-1} \cdot (X - X_i^c)\right) \qquad (3.11)$$

Determining the Combination Factors of the GBFs

The linear combination of GBFs (reduced set) is defined by the factors w_i.

$$\tilde{f}(X) := \sum_{i=1}^{I} w_i \cdot f_i^{Gs}(X) \qquad (3.12)$$

The approach for determining appropriate combination factors is as follows. First, the I basis functions are applied to the J vectors X_j of the training set. This results in a matrix $\mathcal{V}$ of similarity values with J rows and I columns. Second, we define an J-dimensional vector Y comprising the desired output values $y_1, \cdots, y_J$ for the J training vectors. Third, we define a vector W, which comprises the unknown combination factors $w_1, \cdots, w_I$ of the basis functions. Finally, the problem is to solve the equation $\mathcal{V} \cdot W = Y$ for the vector W. According to Press *et al.* [134, pp. 671-675], we compute the pseudo inverse of $\mathcal{V}$ and determine the optimal vector W directly.

$$\mathcal{V}^\dagger := (\mathcal{V}^T \cdot \mathcal{V})^{-1} \cdot \mathcal{V}^T, \qquad W := \mathcal{V}^\dagger \cdot Y \qquad (3.13)$$

The sample data Ω_{GBF} have been defined previously as set of elements, each one consisting of input vector and output scalar. GBF network learning can be generalized in the sense of treating also output vectors instead of output scalars. For this case, we simply compute a specific set of combination factors of the GBFs for each output dimension, respectively.

Functionality of GBF Networks

The use of GBF networks allows a *nonlinear dimension reduction* from input to output vectors (because of the nonlinearity of the Gaussians). Equations (3.11) and (3.12) define an approximation scheme which can be used for relevant functions of object or situation recognition. The approximation scheme is popular in the neural network literature under the term *regularization neural network* [21, pp. 164-191], and we call it *GBF network* to emphasize the Gaussians. The approach does not assume a normal density distribution of the whole population of input vectors. However, it is assumed that the density distribution of input vectors can be approximated by a combination of several normal distributed subpopulations. Actually, this is the motivation for using Gaussians as basis functions in the network. Each GBF must be responsible for a normal distributed subpopulation of input vectors. GBF network learning helps to overcome the serious *bias problem* in high-level machine learning [169] and parameter estimation [134]. The dynamic structure of the network, to be changed and controlled on the basis of error feedback [29], lets the learning method go beyond pure function approximation.

Visualization of GBF Networks

A GBF network consists of an input layer, a layer of hidden nodes and an output layer. The input layer and output layer represent the input and output of the function approximation, the nodes of the hidden layer are assigned to the GBFs (see Figure 3.2).

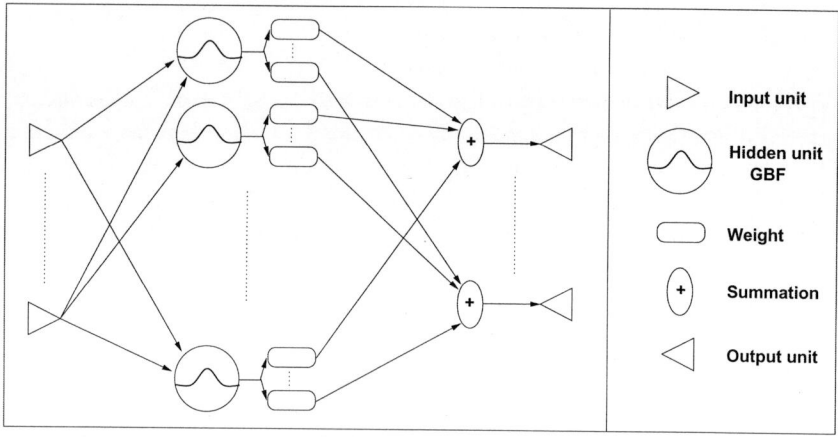

Fig. 3.2. Architecture of a GBF network.

Representing Recognition Functions by GBF Networks

The GBF network defines function $\tilde{f}$ which is an approximation of unknown function f. For evaluating the accuracy of the approximation we must apply testing data consisting of new input vectors and accompanying desired output values. The deviation between computed output and desired output can be formalized by defining a function f^{im}, which plays the role of an implicit function according to equation (3.1).

$$f^{im}(B, Z) := \tilde{f}(X) - y \qquad (3.14)$$

Vector B consists of all parameters and combination factors of the Gaussians, and vector Z consists of input vector X and desired output value y. We obtain a PAC approximation subject to the parameters P^r and ψ, if with a probability of at least P^r the magnitude of $f^{im}(B, Z)$ is less than ψ.

In the following, the paradigm of learning GBF networks is applied for the purpose of object recognition. We assume that an individual GBF network is responsible for each individual object. Concretely, each GBF network must approximate the appearance manifold of the individual object. Based on a training set of appearance patterns of an individual object, we obtain the GBF network which defines function $\tilde{f}$. Without loss of generality, function $\tilde{f}$ can be learned such that each appearance patterns of the object is transformed to the value 1, approximately. Equation (3.14) is specialized to

$$f^{im}(B, Z) := \tilde{f}(X) - 1 \qquad (3.15)$$

By applying equation (3.4) we obtain function $f^{Gi}(B, Z)$ whose values are restricted in the unit interval. The closer the value at the upper bound (value 1) the more reliable a pattern belongs to the object. Several GBF networks have been trained for individual objects, respectively. By putting a new appearance pattern into all GBF networks, we are able to discriminate between different objects based on the maximum response among the networks.

Visualization of GBF Networks Applied to Object Recognition

Let us assume three patterns from a target object (maybe taken under different viewing angles) designated by X_1, X_2, X_3 which are represented in the high-dimensional input (pattern) space as points. In the left diagram of Figure 3.3 these points are visualized as black disks. We define three spherical GBFs centered at these points and make a summation. The result looks like an undulating landscape with three hills, in which the course of the *ridge* goes across the hills. Seen from top view of the landscape the ridge between the hills can be approximated by three straight lines. In the right diagram of Figure 3.3 the result of function f^{Gi} is shown when moving along points of the ridge. The function is defined in equation (3.4) and is based on equation (3.15) specifically.

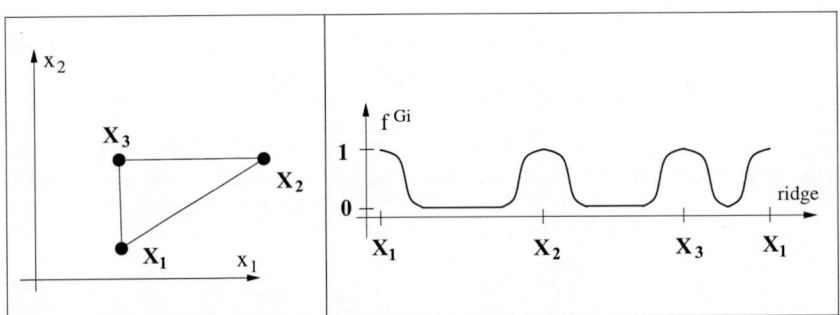

Fig. 3.3. (Left) Input space with three particular points which are positions of three Gaussians, virtual straight lines of the 2D projection of the ridge; (Right) Result of function f^{Gi} along the virtual straight lines between the points.

The manifold of all patterns for which equation (3.1) holds just consists of the target patterns denoted by X_1, X_2, X_3. By accepting small deviations for f^{Gi} from value 1, which is controlled by ζ in equation (3.7), we enlarge the manifold of patterns and thus make a generalization, as shown in Figure 3.4.

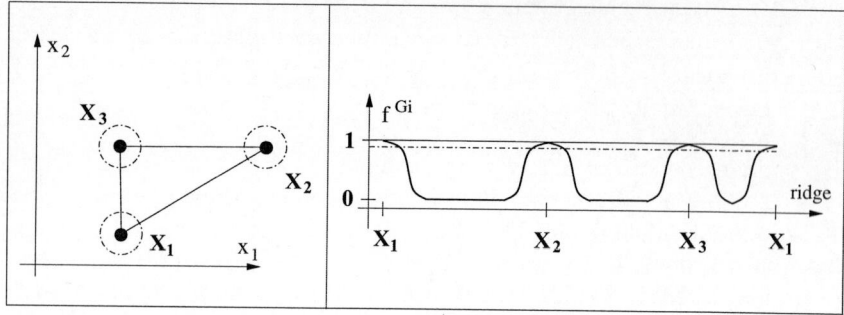

Fig. 3.4. (Left) Input space with three particular points, three Gaussians with small extents; (Right) Result of function f^{Gi} along the virtual straight lines between the points, small set of points of the input space surpasses threshold ζ.

The degree of generalization can be controlled by the extents of the Gaussians via factor τ (see equation (3.10)). An increase of τ makes the Gaussians more flat, with the consequence that a larger manifold of patterns is accepted subject to the same threshold ζ (see Figure 3.5).

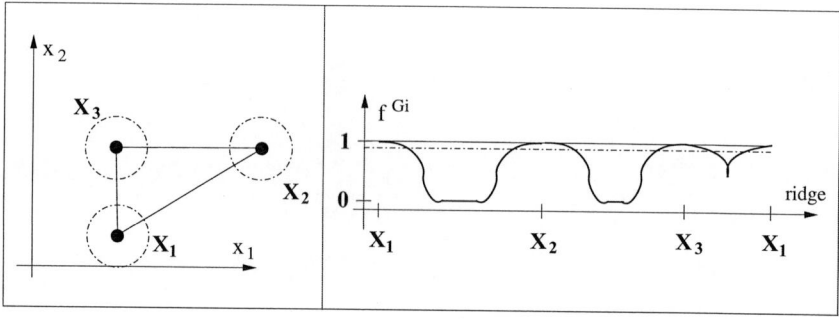

Fig. 3.5. (Left) Input space with three particular points, three Gaussians with large extents; (Right) Result of function f^{Gi} along the virtual straight lines between the points, large set of points of the input space surpasses threshold ζ.

In various applications, to be presented in Section 3.3, we will treat the *overgeneralization/overfitting dilemma*.

3.2.3 Canonical Frames with Principal Component Analysis

For an appearance-based approach to recognition the input space is high-dimensional as it consists of large-scaled patterns, typically. Due to the high-dimensional input space, the training of recognition operators is time-consuming. For example, the clustering step in learning an appropriate GBF network is time-consuming, because frequent computations of similarity values between high-dimensional training vectors are necessary. This subsection introduces the use of *canoncical frames* which will be combined with GBF networks in Subsection 3.4.4. The purpose is to obtain reliable recognition operators which can be learned and applied efficiently.

Role of Seed Images for Recognition

The increased efficiency and reliability is based on a pre-processing step prior to learning and application. It makes use of the following ground truth. Images taken under similar conditions, *e.g.* similar view angles, similar view distance, or similar illumination, contain diverse *correlations* between each other. In the sense of *information theory*, for similar imaging conditions the entropy of taking an image is low, and for a significant change of the imaging conditions the entropy is high. Accordingly, for the purpose of object or situation recognition it makes sense to take a small set of important images (we call them *seed images*, *seed apperances*, or *seed patterns*) and determine a low-dimensional sub-space thereof. Any new image is supposed to be in high correlation with one of the seed images, and therefore it is reasonable to approximate it in the low-dimensional sub-space. The set of basis vectors spanning this sub-space is called a *canonical frame (CF)*, which is aligned to the patterns resulting from the seed images of the object. From the mathematical point of view, a canonical frame is a coordinate system, however, the coordinate axes represent feature vectors. Coming back to the beginning of this subsection, the pre-processing step consists in mapping new views into the canonical frame.

Interplay of Implicit Function and Canonical Frame

The implicit function in equation (3.1), which is responsible for approximating a manifold of view patterns, will be represented in the canonical frame. We impose three requirements. First, in the canonical frame the implicit function should have a simpler description than in the original frame. For example, in the canonical frame a hyper-ellipsoid would be in normal form, *i.e.* the center of the ellipsoid is in the origin of the frame, and the principal axes are collinear with the frame axes. The dimension of parameter vector B is lower compared to the one in the original frame. Second, in the canonical frame the equation (3.1) must hold perfectly for all seed patterns which are represented as vector Z, respectively. In this case, the parameters in vector B are invariant features of the set of all seed patterns. Third, the implicit function should consider generalization principles as treated in the paradigms of Machine Learning [177, pp. 349-363]. For example, according to the *enlarge-set rule* and the *close-interval rule*, the implicit function must respond continuous around the seed vectors and must respond nearly invariant along certain courses between successive seed vectors (in the pattern space). For avoiding hazardous decisions, which are caused by over-generalizations, the degree of generalization should be low.

An appropriate canonical frame together with an implicit function can be constructed by principal component analysis. The usual application of PCA relies on a large set of multi-dimensional, perverably Gaussian-distributed, vectors. It computes a small set of basis vectors spanning a lower-dimensional sub-space which include rather accurate approximations of the original vectors. However, we apply PCA to the small set of seed patterns and only move the original frame into another one which contains the seed vectors exactly. No effective dimension reduction is involved because all seed patterns are equal significant. Taking the covariance matrix of the seed patterns into account, we use the normalized eigenvectors as basis vectors of unit length. The representation of a seed pattern in the canonical frame is by Karhunen-Loéve expansion. Implicit function f^{im} is defined as a hyper-ellipsoid in normal form with the half-lengths of the ellipsoid axes defined dependent on the eigenvalues of the covariance matrix, respectively. As a result, the seed vectors are located on the orbit of this hyper-ellipsoid, and invariants are based on the half-lengths of the ellipsoid axes.

Principal Component Analysis for Seed Patterns

Let $\Omega_{PCA} := \{X_i | X_i \in R^m; i = 1, \cdots, I\}$ be the vectors representing the seed patterns of an object. Taking the mean vector X^c we compute the matrix

$$\mathcal{M} := (X_1 - X^c, \cdots, X_I - X^c) \tag{3.16}$$

It is easy proven that the *covariance matrix* is obtained by

$$\mathcal{C} := \frac{1}{I} \cdot \mathcal{M} \cdot \mathcal{M}^T \tag{3.17}$$

We obtain the eigenvectors $E_1, \cdots, E_I$ of the covariance matrix $\mathcal{C}$. The I vectors $X_1, \cdots, X_I$ of Ω_{PCA} can be represented in a coordinate system which is defined by just $(I - 1)$ eigenvectors and the origin of the system. A simple example for $I = 2$ shows the principle. The difference vector $(X_1 - X_2)$ is obtained as the first eigenvector E_1 and is used as axis of a one-dimensional coordinate system. Additionally, the mean vector X^c is taken as origin. Then, the two vectors X_1 and X_2 are located on the axis with certain coordinates. A second axis is not necessary, and in the one-dimensional coordinate system each of the two vectors has just one coordinate, respectively. This principle can be generalized to the case of I vectors $X_1, \cdots, X_I$ which is represented in a coordinate system of just $(I - 1)$ axes, and just $(I - 1)$ coordinates are needed for obtaining a unique location.

Consequently, the principal component analysis must yield at least one eigenvalue equal to 0 and therefore the number of relevant eigenvalues $\lambda_1, \cdots, \lambda_{I-1}$ is at most $(I - 1)$. The vectors of Ω_{PCA} represent seed patterns which are hardly correlated with each other. It depends on this degree of non-correlation which determines the actual number of relevant eigenvectors, e.g. less or equal to $(I - 1)$.

Let us assume the relevant eigenvectors $E_1, \cdots, E_{I-1}$. The Karhunen-Loéve expansion of a vector X, i.e. the projection of the vector into the $(I - 1)$-dimensional eigenspace, is defined by

$$\hat{X} := (\hat{x}_1, \cdots, \hat{x}_{I-1})^T := (E_1, \cdots, E_{I-1})^T \cdot (X - X^c) \tag{3.18}$$

Ellipsoidal Implicit Functions Based on PCA

Based on the eigenvalues of the PCA and the Karhunen-Loéve expansion, we introduce the following implicit function which defines a hyper-ellipsoid.

$$f^{im}(B, Z) := \left(\sum_{l=1}^{I-1} \frac{\hat{x}_l^2}{\kappa_l^2} \right) - 1 \tag{3.19}$$

Notice the close relationship to the special case of a 2D ellipse formula in equation (3.2). Input-output vector $Z := \hat{X} := (\hat{x}_1, \cdots, \hat{x}_{I-1})^T$ is defined according to equation (3.18). Parameter vector $B := (\kappa_1, \cdots, \kappa_{I-1})^T$ contains the parameters κ_l denoting the half-lengths of the ellipsoid axes in normal form. We define these parameters as

$$\kappa_l := \sqrt{(I - 1) \cdot \lambda_l} \tag{3.20}$$

Let vectors $\hat{X}_1, \cdots, \hat{X}_I$ be the KLE of the seed vectors $X_1, \cdots, X_I$, as defined in equation (3.18). For the special cases of assigning these KLE-transformed seed vectors to Z, respectively, we made the following important observation. By taking equation (3.19) into account, the equation (3.1) holds for all seed vectors, perfectly. That is, all seed vectors have a particular location in the canonical frame, namely, they are located on the orbit of the defined hyper-ellipsoid.[4] The hyper-ellipsoid is an *invariant description* for the set of seed vectors.

The PCA determines a $(I - 1)$-dimensional hyper-ellipsoid based on a set of I seed vectors, and all these are located on the orbit of the ellipsoid. Generally, more than I points are necessary for fitting a unique $(I - 1)$-dimensional hyper-ellipsoid. The question of interest is, which hyper-ellipsoid will be determined by the PCA. It is well known that PCA determines the first principal axis by maximizing the variances which are obtained by an orthogonal projection of the sample points on hypothetical axis, respectively. Actually, this is the bias which makes the fitting unique.

For illustration, we describe the 2D case of constructing ellipses. It is well-known that 5 points are necessary for fitting a unique 2-dimensional ellipse, however, only 3 points are available. Figure 3.6 shows two examples of ellipses, each fitting the same set of three points. The ellipse in the left image has been determined by PCA, and ellipse in the right image has been

---

[4] The proof is given in Appendix 1.

fitted manually. In the figure, the projection of the sample points on two
hypothetical axes is shown. The variance on the right is lower than on the
left, as expected, because the variance on the left is the maximum.

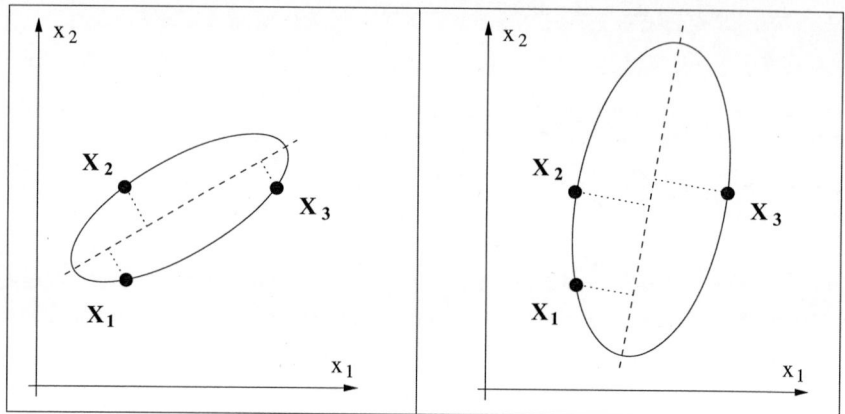

Fig. 3.6. Ellipses fitted through three points; (Left) Ellipse determined by PCA,
showing first principal axis, determined by maximizing the variance; (Right) Ellipse
determined manually with less variance along the first principal axis.

Motivation for Ellipsoidal Implicit Functions

A nice property can be observed concerning the aspect of generalization. The
set of seed patterns is just a subset of the manifold of all patterns for which
equation (3.1) holds. Actually, the size of this manifold, *e.g.* the perimeter of
a 2D ellipse, correlates with the degree of generalization. PCA produces mod-
erate generalizations by avoiding large ellipsoids. The ellipsoid determined by
PCA can be regarded as a compact grouping of the seed vectors. This is also
observed in Figure 3.6, which shows that the left ellipse (produced by PCA)
is smaller than the right one.

The ellipsoidal implicit function considers the enlarge-set and the close-
interval rule as requested by learning paradigms. In the following, this will
be demonstrated visually. Coming back to the problem of object recognition,
the question is, which views others than the seed views can be recognized
via the implicit function. Let us assume three patterns from a target object
denoted by X_1, X_2, X_3 which are represented in the high-dimensional input
(pattern) space as points. In the left diagram of Figure 3.7 these points are
visualized as black disks (have already been shown in Figure 3.6). We determine the two-dimensional eigenspace and the eigenvalues by PCA, and
construct the relevant 2D ellipse through the points. The right diagram of
Figure 3.7 shows a constant value 1 when applying function f^{Gi} (as defined
in equations (3.19) and (3.4)) to all orbit points of the ellipse. Therefore the
generalization comprises all patterns on the ellipse (close-interval rule).

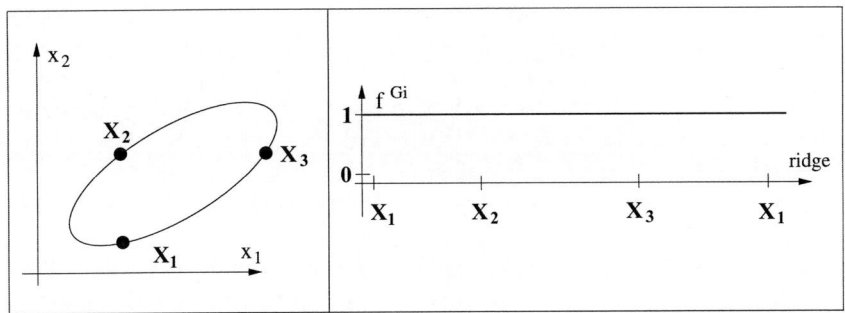

Fig. 3.7. (Left) Input space with three particular points from which a 2D ellipse is defined by PCA; (Right) Result of function f^{Gi} along the ellipse, constant 1.

For a comparison with the GBF network approach, one can see the relevant curve in the right diagram of Figure 3.3. There, a generalization did not take place, *i.e.* equation (3.1) just holds for the Gaussian center vectors. In the PCA approach the degree of generalization can be increased furthermore by considering the threshold ζ and accepting small deviations for f^{Gi} from 1. The relevant manifold of patterns is enlarged, as shown by the dotted band around the ellipse in Figure 3.8 (enlarge-set rule).

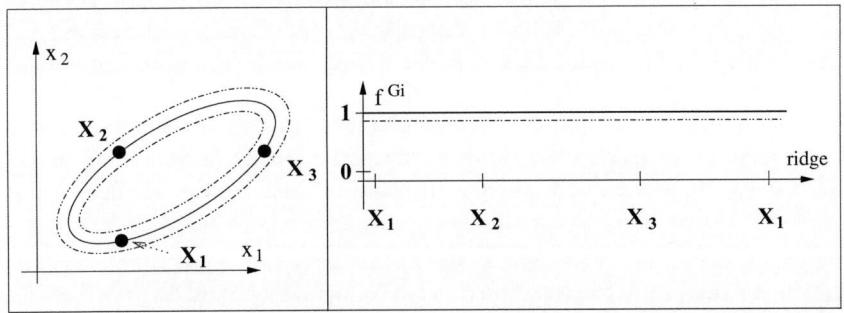

Fig. 3.8. (Left) Input space with three particular points from which a 2D ellipse is defined by PCA, small deviations from this ellipse are constrained by an inner and an outer ellipse; (Right) Result of function f^{Gi} along the ellipse, which is constant 1, accepted deviations are indicated by horizontal lines with offset $\mp\zeta$.

Discussion of GBF Approach and PCA Approach

The following résumé finishes this section. In the GBF approach, the generalization is flexible, because the extent of each Gaussian can be controlled individually. In comparison to this, the generalization in the PCA approach is less fexible, because the manifold of patterns is enlarged globally by taking patterns around the whole ellipse orbit into account. In the GBF approach, it is expected that the sample data can be separated such that the distribution of the elements in each cluster follows a normal probability density. The PCA approach uses a small set of seed vectors (which are not normal distributed) for constructing a canonical frame of the appearance patterns of an object, and thus reduces the high-dimensional input space significantly. The elliptical orbit obtained by PCA describes a global relationship between the seed patterns.

In the next Section 3.3, we train GBF networks for the recognition of objects and the scoring of situations. Different configurations and parameterizations of GBF networks affect the balance between invariance and discriminability of the recognition function. Finally, Section 3.4 combines the GBF and PCA approaches according to a strategy such that the advantages of each are exploited. The main purpose is to obtain a recognition function which obeys an acceptable compromise between invariance, discriminability, and efficiency.

3.3 GBF Networks for Approximation of Recognition Functions

In this section, GBF networks are applied for globally representing a manifold of different object appearances (due to changes of viewing conditions) or a manifold of different grasping situations (due to movements of a robot gripper). Just a small set of distinguished appearance patterns are taken from which to approximate the relevant appearance manifold roughly.[5] In the spirit of applying Occam's razor to object or situation recognition, the sparse effort for training and representation may be sufficient in the actual application, *i.e.* reach a certain quality of PAC-recognition.[6] The purpose of visual demonstration and experimentation (with several configurations of GBF networks) is to clarify this issue.

3.3.1 Approach of GBF Network Learning for Recognition

Object recognition has to be grounded on features which discriminate between the target object and other objects or situations and can be extracted from the image easily. In our approach, an object is recognized in a certain image area by applying a learned recognition function to the signal structure of this area. The output of the *recognition function* should be a real value between 0 and 1, which encodes the confidence, that the target object is depicted in the image area. Regardless of the different appearance patterns of the object the recognition function should compute values near to 1. On

---

[5] Approaches for 3D object recognition which use a small set of privileged or canonical views are known as *aspect graph methods* [137].

[6] Poggio and Girosi introduced the concept of *sparsity in approximation* techniques [131, 64].

the other hand, the recognition function should compute values near to 0 for image areas depicting any counter object or situation.

Alternatively, in certain applications, we need a scoring function for a more fine-grained evaluation of situations. This is typically the case in robotic servoing processes which deal with continually changing situations (see Chapter 4). For example, for grasping a target object appropriately, a certain function must be responsible for evaluating the grasping stability while approaching the object [44]. Based on this, the manipulator is servoed to the most stable grasping pose in order to grasp the target object.

Constructing GBF Networks for Recognition

GBF networks are used for learning the recognition or scoring function. In case of object recognition, we must acquire distinguished samples of the appearance manifold of the target object by changing the imaging conditions systematically and taking a discrete set of images. In case of situation evaluation, we must acquire distinguished samples of intermediate situations along with the desired scores. Optionally, we transform the relevant image patch with specific filters in order to enhance certain properties, or to simplify the complexity of the pattern manifold. According to the approach for learning a GBF network, the generated set of training patterns must be clustered with regard to similarity by taking the required quality of PAC-recognition into account (*e.g.* using the error-based ISODATA algorithm mentioned in Subsection 3.2.2). Depending on the actual requirements and on the specific strategy of image acquisition, maybe the clustering step can be suppressed if each distinguished sample plays a significant role of its own, *i.e.* each distinguished sample represents a one-element cluster. For each cluster a GBF is defined, with the mean vector used as Gaussian center vector, and the covariance matrix computed from the distribution of vectors in the cluster (for a one-element cluster the identity matrix is used as covariance matrix). The extents of the GBFs, defined by the combination of the covariance matrix C and a factor τ are responsible for the generalizing ability (see equation (3.11)), *i.e.* usefulness of the operator for new patterns (not included in the training set). Factor τ for controlling the Gaussian overlap is subject of the experiments (see later on). The final step of the learning procedure is to determine appropriate combination factors of the GBFs by least squares fitting (*e.g.* using the *pseudo inverse technique* mentioned in Subsection 3.2.2 or singular value decomposition).

Appearance-Based Approach to Recognition

The learned GBF network represents a recognition or scoring function. The application to a new pattern is as follows. The input nodes of the GBF network represent the input pattern of the recognition function. The hidden nodes are defined by I basis functions, and all these are applied to the input pattern. This hidden layer approximates the appearance manifold of the target object or a course of situations. The output node computes the recognition or scoring value by a weighted combination of results coming from the basis functions. The input space of the GBF network is the set of all possible patterns of the pre-defined size, but each hidden node responds significantly only for a certain subset of these patterns. Unlike simple applications of GBF networks, in this application of object recognition or situation scoring, the dimension of the input space is extremely high (equal to the pattern size of the target object, *e.g.* $15 \times 15 = 225$ pixel). The high-dimensional input space is projected nonlineary into a one-dimensional output space of confidence values (for object recognition) or scoring values (for situation evaluation), respectively.

Different Configurations/Parametrizations of GBF Networks

By carefully spreading and parameterizing the Gaussian basis functions, an optimal PAC operator can be learned, which carries out a compromise between the invariance and discriminability criterion. The invariance criterion strives for an operator, which responds nearly equal for any appearance pattern of the target object. The discriminability criterion aims at an operator, which clearly discriminates between the target object and any other object or situation. This conflict is also known under the terms overgeneralization versus overfitting. On account of applying the principle of minimum description length to the configuration of GBF networks, it is desirable to discover the minimum number of basis functions to reach a required quality of PAC function approximation. In the experiments of the following subsections we show the relationship between number and extents of the GBFs on the one hand and the invariance/discriminability conflict on the other hand. The following interesting question must be clarified by the experiments.

> How many GBFs are needed and which Gaussian extents are appropriate to reach a critical quality of PAC-recognition ?

3.3.2 Object Recognition under Arbitrary View Angle

For learning an appropriate operator, we must take sample images of the target object under several view angles. We rotate the object by using a rotary table and acquire orientation-dependent appearance patterns (size of the object patterns $15 \times 15 = 225$ pixel). Figure 3.9 shows a subset of eight patterns from an overall collection of 32. The collection is devided into a training and a testing set comprising 16 patterns each. The training set has been taken by equidistant turning angles of $22.5°$ degrees, and the testing set differs by an offset of $10°$ degrees. Therefore, both in the training and testing set the orientation of the object varies in discrete steps over the range of $360°$ degrees.

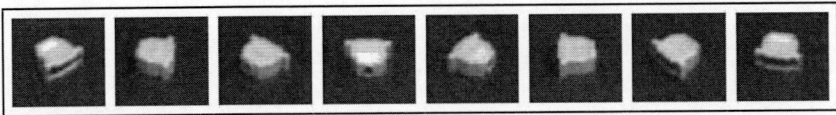

Fig. 3.9. The target object is shown under eight equidistant rotation angles. The patterns are used to learn an operator for object recognition under arbitrary view angle.

The collection of GBFs and their combination factors are learned according to the approach of Subsection 3.3.1. By modifying the number and/or the extent of the GBFs, we obtain specific GBF network operators.

Experiments to Object Recognition under Arbitrary View Angle

In the first experiment, a small extent has been chosen, which implies a spare overlap of the GBFs. By choosing $2, 4, 8$, and 16 GBFs, respectively, four variants of GBF networks are defined to recognize the target object. Figure 3.10 shows the four accompanying curves (a), (b), (c), (d) of confidence values which are computed by applying the GBF networks to the target object of the test images. The more GBFs are used, the higher the confidence values for recognizing the target. The confidence values vary significantly when rotating the object, and hence the operators are hardly invariant.

The second experiment differs from the first in that a large extent of the GBFs has been used, which implies a broad overlap. Figure 3.11 shows four curves of confidence values, which are produced by the new operators. The invariance criterion improves and the confidence nearly takes the desired value 1. Taking only the *invariance aspect* into account, the operator characterized by many GBFs and large extent is the best (curve (d)).

The third experiment incorporates the *discriminability criterion* into object recognition. An operator is discriminable, if the recognition value computed for the target object is significant higher than those of other objects. In the experiment, we apply the operators to the target object and to three test objects (outlined in Figure 3.12 by white rectangles). Based on 16 GBFs we systematical increase the extent in 6 steps. Figure 3.13 shows four curves related to the target object and the three test objects. If we enlarge the extent of the GBFs and apply the operators to the target object, then a slight increase of the confidence values occurs (curve (a)). If we enlarge the extent in the same way and apply the operators to the test objects, then the confidence values increase dramatically (curves (b), (c), (d)). Consequently, the curves for the test objects approach the curve for the target object. An increase of the extent of the GBFs makes the operator more and more unreliable. However, according to the previous experiment an increasing extent makes the operator more and more invariant with regard to object orientation. Hence, a compromise has to be made in specifying an operator for object recognition.

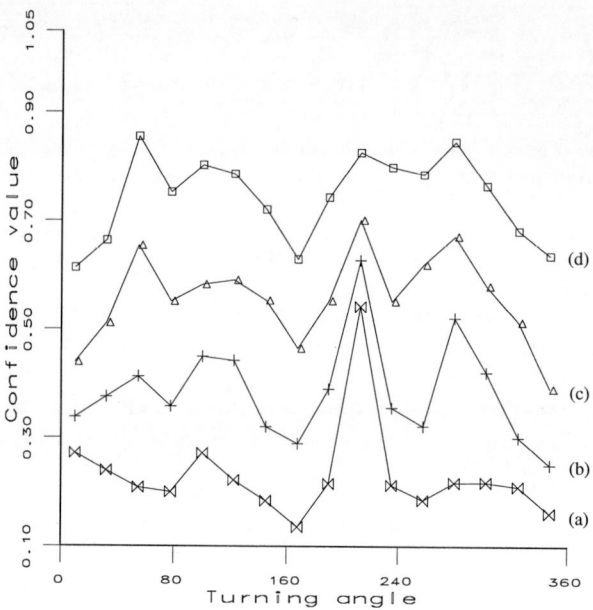

Fig. 3.10. Different GBF networks are tested for object recognition under arbitrary view angle. The network output is a confidence value, that a certain image patch contains the object. The curves (a), (b), (c), (d) show the results under changing view angle using networks of $2, 4, 8, 16$ GBFs, respectively. The more GBFs, the higher the confidence value. Due to a *small GBF extent* the operators are *not invariant* under changing views.

Aspects of PAC Requirements in Recognition For this purpose we formulate PAC requirements (see Definition 3.1 of PAC-recognition). First, a probability threshold P^r is pre-specified which serves as a description of the required quality of the operator, *e.g.* $P^r := 0.9$. Second, the maximum value for a threshold ζ_1 is determined, such that with the probability of at least P^r the confidence value of a target pattern surpasses ζ_1. Third, the minimum value for a threshold ζ_2 is determined, such that with the probability of at least P^r the confidence value of a counter pattern falls below ζ_2. Finally, if ζ_1 is less than ζ_2, then we define ζ as the mean value between both. In this case, the recognition function is PAC-learned subject to the parameters P^r and ζ. Otherwise, the recognition function can not be PAC-learned subject to parameter P^r. In this case, another configuration and/or parameterization of a GBF network have to be determined. Actually, the ideal GBF network would be the one which provides the highest value for probability threshold P^r. A sophisticated approach is desirable which has to optimize all unknowns of a GBF network in combination.[7]

---

[7] We do not focus on this problem, instead refer to Orr [122].

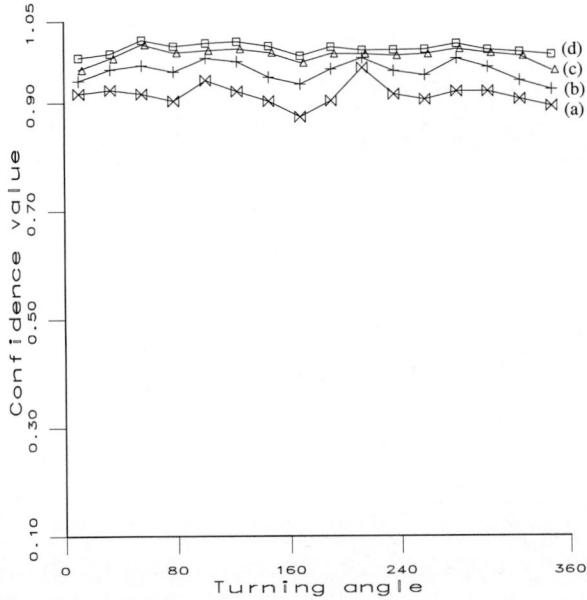

Fig. 3.11. Similar experiments like the one in Figure 3.10. However, a *large extent* of the GBFs has been used. The learned operators respond *nearly invariant* under varying view angles.

200 200

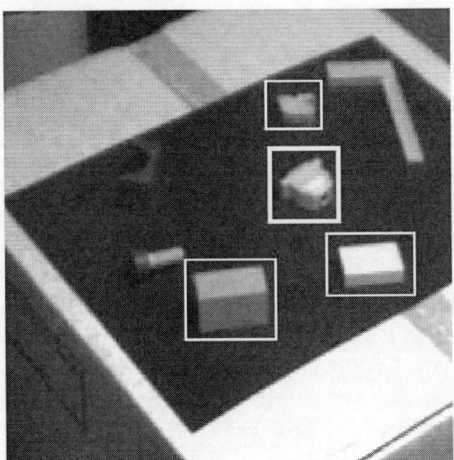

Fig. 3.12. The image shows a certain view of the target object (in a bold rectangle) and three test objects (in fine rectangles). The GBF network for object recognition should detect the target object in this set of four candidates.

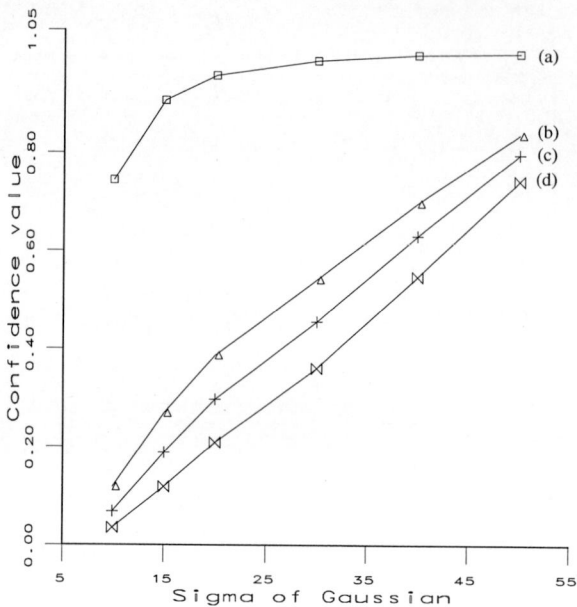

Fig. 3.13. Six GBF networks have been constructed each with equal GBF number, but with different GBF extents. Each GBF network has been applied to the image patch of the target object and to the patches of the three test objects. The GBF network computes a confidence value, that the patch contains the target object. The curves show the confidence values versus the extents of the GBFs. The target object (curve (a)) can be discriminated from the test objects (curves (b),(c),(d)) quite good by GBF networks of small extents. However, for larger extents the discriminating power decreases.

It has to be mentioned that in our application of the PAC methodology a learned operator for object recognition has been validated only for a limited set of test examples, but of course not for the infinite set of all possible situations. Actually, the probability threshold P^r is treated as a frequency threshold. In consequence of this, there is no guarantee that the obtained quality will hold also for other possible imaging conditions in this application scenario. However, our approach of quality assessment is the best practical way to proceed, *i.e.* obtaining a useful threshold P^r. The designer must provide reasonable training and test scenarios such that the quality estimations of the learned operators will prove reliable.[8]

---

[8] Problems like these are treated with the approach of *structural risk minimization* which actually is a minimization of the sum of empirical risk and the so-called *VC-confidence* [31]. However, in our work we introduce another approach of dealing with ambiguous situations, which is based on gathering additional information from the images. The relationship between objects and cameras can be changed advantageous, *e.g.* by continual feedback control of robot manipu-

3.3.3 Object Recognition for Arbitrary View Distance

In this subsection similar experiments are carried out for object recognition under arbitrary view distance. In order to learn an appropriate operator, we must take sample images of the target object under several spatial distances between object and camera. Figure 3.14 shows on the left the image of a scene with the target object and other objects taken under a typical object–camera distance. On the right, a collection of 11 training patterns depicts the target object, which has been taken under a systematic decrease of the camera focal length in 11 steps. The effect is similar to decreasing the object–camera distance. The size of the object pattern changes from 15×15 pixel to 65×65 pixel. We define for each training pattern a single GBF (*i.e.* avoiding clustering), because each pattern encodes essential information. The combination factors of the GBFs are determined as before. A further collection of 10 test images has been acquired, which differs from the training set by using intermediate values of the camera focal length.

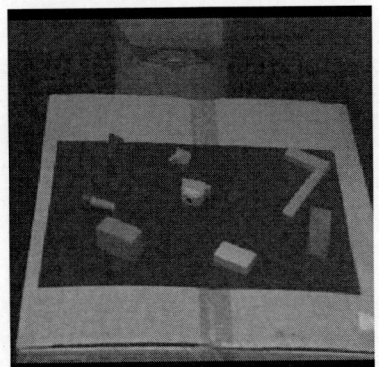 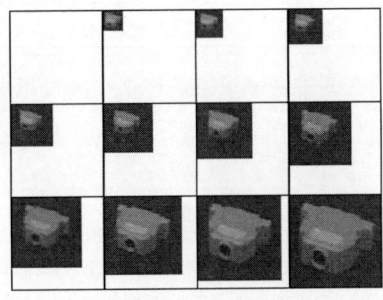

Fig. 3.14. On the left, an image of a whole scene has been taken including the target object. On the right, a collection of 11 images is taken just from the target object under systematic increase of the inverse focal length. The effect is similar to decreasing the object–camera distance. This collection of images is used to learn an operator for object recognition under arbitrary view distance.

Experiments to Object Recognition under Arbitrary View Distance

We constructed three operators for object recognition by taking small, middle, and large extent of the GBFs (Figure 3.15). In the first experiment, these operators have been applied to the target object of the test images. In curve (a) the confidence values are shown for recognizing the target object by taking a small extent into account. The confidence value differs significantly

---

lator or head (see Subsection 4.3.6 later on), for obtaining more reliable and/or informative views.

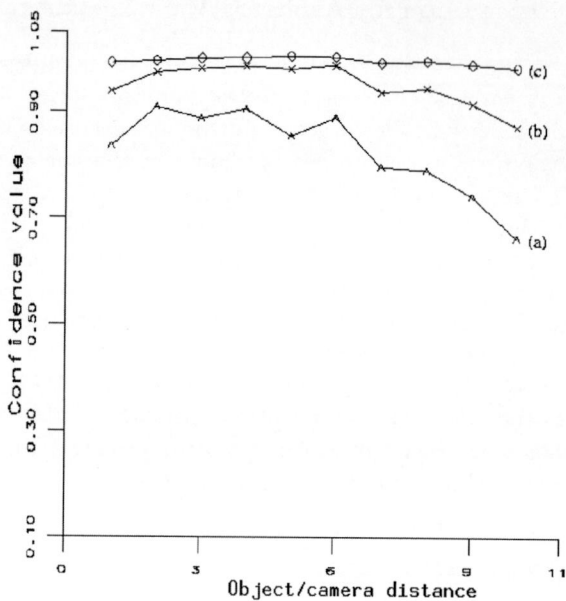

Fig. 3.15. Three GBF networks are tested each with equal GBF number, but differing by small, middle, and large GBF extent. Each network is applied to the target object in 10 test images, which differ from each other in the size of the depicted object, *i.e.* in the view distance. The network output gives a confidence value, that the image patch contains the target object. For small or middle GBF extents (curves (a), (b)) the learned operators are hardly invariant under changing view distance. For a large extent (curve (c)) an invariance is reached.

when changing the object–camera distance and is far away from the desired value 1. Alternatively, if we use a middle extent value, then the confidence values approach to 1 and the smoothness of the curve is improved (curve (b)). Finally, the use of a large extent value will lead to approximately constant recognition values close to 1 (curve (c)).

In the second experiment, we investigate the discriminability criterion for the three operators from above. The operators are applied to all objects of the test image (image on the left in Figure 3.14), and the highest confidence value of recognition has to be selected. Of course, it is expected to obtain the highest recognition value from the target object. For comparison, Figure 3.16 depicts once again the confidence values of applying the three operators to the target object (curves (a), (b), (c), equal to Figure 3.15)).

If we apply the operator with large extent value to all objects of the test images, then we obtain higher confidence values frequently for objects other than the target object (see curve (c1)). In those cases, the operator fails to localize the target object. Alternatively, the operator with middle extent values fulfills the discriminability criterion better (curve (b1) surpasses curve

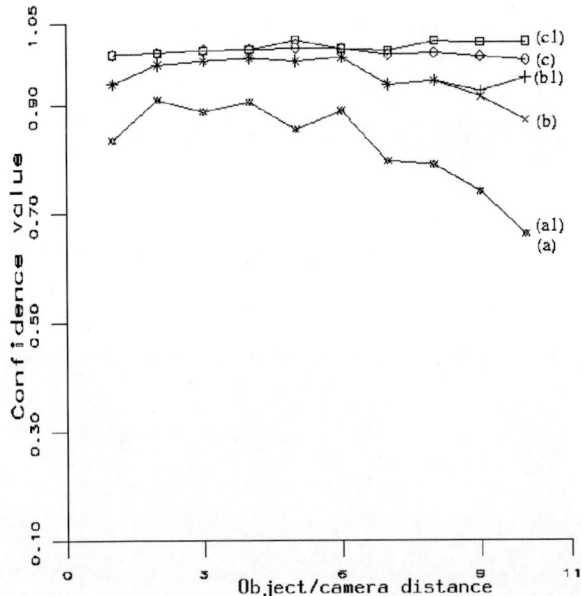

Fig. 3.16. The curves (a), (b), (c) of Figure 3.15 are shown, which are the output of three GBF networks (differing by the extent), when applied just to the patch of the target object under varying view distance. In order to consider the reliability of these values for discriminating target and other objects the three GBF networks has been applied further to the patches of other objects under varying view distance. The left image of Figure 3.14 shows all these objects under a certain view distance. Each GBF network computes for each object patch an output value and the maximum of these values is taken. Repeating this procedure for all three GBF networks and for all view distances yield the curves (a1), (b1), (c1). For a small GBF extent, the curves (a) and (a1) are equal, for a middle extent the curve (b1) surpasses curve (b) sometimes. For a large extent the curve (c1) surpasses curve (c) quite often. Generally, the higher the GBF extent the less reliable the GBF network for object recognition.

(b)) rarely. Finally, the operator with small extent values localizes the target object in all test images. The highest confidence values are computed just for the target object (curve (a) and curve (a1) are identical). Notice again the invariance/discriminability conflict which has to be resolved in the spirit of the previous section.

3.3.4 Scoring of Grasping Situations

So far, we have demonstrated the use of GBF networks for object recognition. Alternatively, the approach is well-suited for the scoring of situations, which describe spatial relations between objects. We will exemplary illustrate specific operators for evaluating grasping situations. A grasping situation is defined to be most stable, if the target object is located between the fingers entirely. Figure 3.17 shows three images, each depicting a target object, two bended grasping fingers, and some other objects. On the left and the right, the grasping situation is unstable, because the horizontal part of the two parallel fingers is behind and in front of the target object, respectively. The grasping situation in the middle image is most stable. For learning to recognize grasping stability, we moved the robot fingers step by step to the most stable situation and step by step moved off afterwards. The movement is photographed in 25 discrete steps. Every second image is used for training and the images in between for testing.

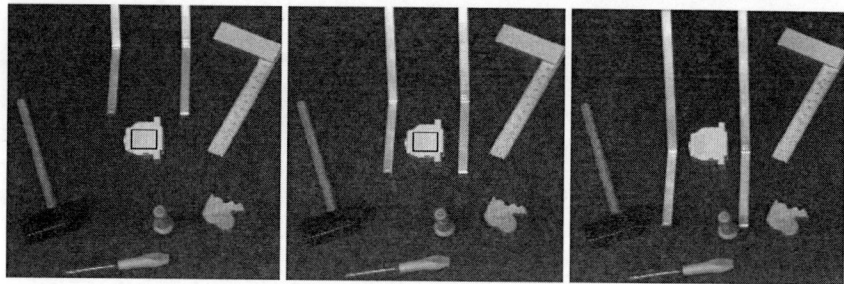

Fig. 3.17. Three typical images of grasping situations are shown. The left and the right grasping situations are unstable, the grasping situation in the middle is stable. Altogether, a sequence of 13 training images is used, which depict first the approaching of the gripper to the most stable grasping situation and then the departure from it. This image sequence is used to learn GBF networks for evaluating the stability of grasping situations.

Using Filter Response Patterns instead of Appearance Patterns

For learning operators, it would be possible to acquire large appearance patterns containing not only the target object, but also certain parts of the grasping fingers. However, the efficiency of recognition decreases if large-sized patterns are used. A filter is needed for collecting signal structure from a large environment into a small image patch. For this purpose, Pauli *et al.* [127] proposed a product combination of two orthogonal directed *Gabor wavelet functions* [138]. By applying such a filter to the left and the middle image in Figure 3.17, and selecting the response of the (black) outlined rectangular area, we obtain the overlay of two response patterns, as shown in Figure 3.18.

A specific relation between grasping fingers and target object results in a specific filter response. Based on filter response patterns, a GBF network can be learned for scoring situations. The desired operator should compute a smooth *parabolic curve* of stability values for the course of 25 grasping situations. For the experiment, we specified many operators by taking different numbers and/or extents of GBFs into account. Figure 3.19 shows the course

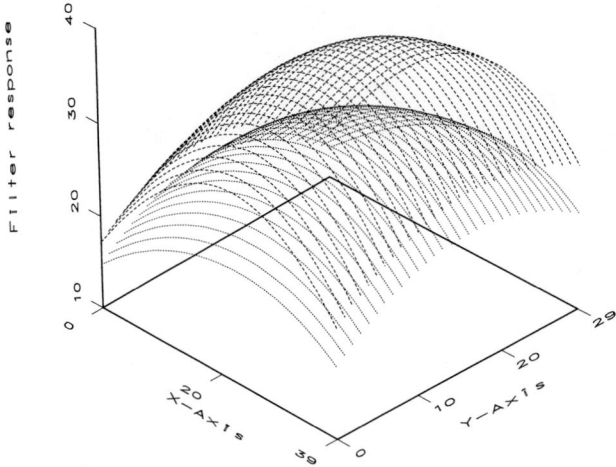

Fig. 3.18. The product combination of two orthogonal directed Gabor wavelet functions can be applied to the image patch of grasping situations. This filter responds specifically to certain relations between target object and grasping fingers. The overlay of the filter response patterns for two different grasping situations are shown. According to this, we can represent the finger–object relation by filter responses and avoid the difficult extraction of symbolic features.

of stability values for two operators. The best approximation can be reached using a large number and large extent of GBFs (see curve (b)).

As a résumé, we conclude that different configurations and parameterizations of GBF networks affect the balance between invariance and discriminability of the recognition or scoring function. The main goal is to obtain a recognition function which obeys an acceptable compromise which considers also the aspect of efficiency. For this purpose we combine the GBF and PCA approaches appropriately.

3.4 Sophisticated Manifold Approximation for Robust Recognition

We introduce a *coarse-to-fine strategy of learning* object recognition, in which a global, sparse approximation of a recognition function is fine-tuned on the basis of space-time correlations and of critical counter situations. This will be done by combining PCA with GBFs such that the advantages of both approaches are exploited. Furthermore, the technique of *log-polar transfor-*

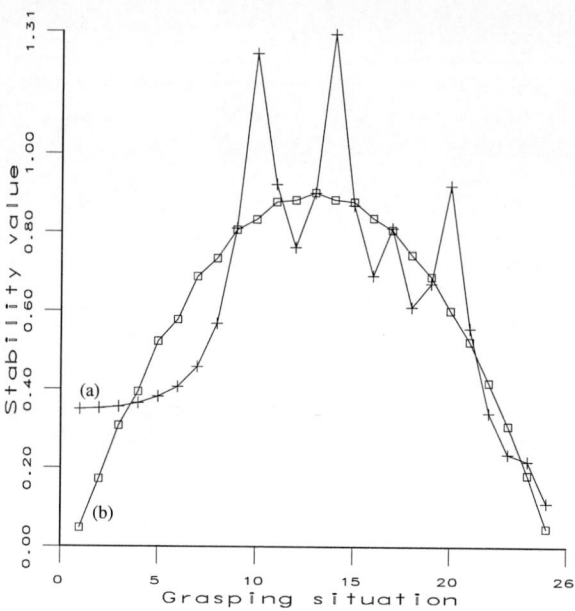

Fig. 3.19. Based on a sequence of 13 training images (which contain the approaching to and the departure from the target object), two GBF networks have been learned. They mainly differ by a small and high number of GBFs, respectively, *i.e.* from the 13 grasping situations a small and high number of clusters are constructed respectively. This image sequence is used for learning a parabolic curve of grasping stability where the maximum should be reached for the middle image of the sequence. Then each GBF network is applied to a succession of 25 different grasping situations depicting once again the approaching and departure. The images include both the 13 training situations and 12 test situations. If using a network with small GBF number, then the resulting course (a) of grasping stability is not the desired one. However, the course (b) resulting from the network with high GBF number is a good approximation of the desired parabolic curve. It can be used for appropriate evaluating grasping situations.

mation will be applied for reducing the manifold complexity in order to obtain an efficient recognition function.

3.4.1 Making Manifold Approximation Tractable

The purpose of recognition is to distinguish the pattern manifold of the target object from the manifolds of counter objects or counter situations. The invariance characteristic of a recognition function makes sense only if a certain level of discriminability is included. This aspect must be considered more directly in a strategy of acquiring a recognition function. The complexity of the boundary between target and counter manifolds is directly correlated with the complexity of the recognition function, *i.e.* its description length. However, a recognition function with large description length is non-efficient in application, which can be realized exemplary for the case of using GBF networks. The center patterns of all GBFs must be compared with the input pattern, and therefore, both the size of patterns and the number of GBFs affect the complexity of the recognition function. In the previous Section 3.3, we acquired GBF networks exclusively for object recognition under different view angles or different view distances. If both variabilities have to be considered in common, possibly including furthermore a variable illumination, then the complexity of the pattern manifold increases significantly, and in consequence of this, much more GBFs are needed for obtaining an appropriate GBF network. This raises also the issue of how much images to take under which conditions. It is desirable to keep the effort of visual demonstration as low as possible and generalize appropriately from a small set of examples.

> In summary, the recognition function should reach a certain level of robustness and both acquisition and application should be efficient.

Object recognition and situation scoring are embedded in a camera-equipped robot system in which the agility of cameras can be exploited to execute a visual inspection task or a vision-supported robotic task. This aspect is the driving force to work out concepts of solution for the conglomerate of problems presented above.

Constraining the Possible Relationships

Depending on the specific task and the specific corporeality of the robot, the *possible relationships* between objects and cameras are constrained. *E.g.*, a camera fastened on a robot arm can move only in a restricted working space, and, therefore, the set of view angles and the set of view distances relative to a fixed object is restricted. The complexity of the pattern manifold decreases by considering in the training process only the relevant appearances.

The relation between objects and cameras changes dynamically. *E.g.*, a manipulation task is solved by continuous control based on visual information. The complexity of the *recognition or scoring function decreases* by putting/keeping the camera in a specific spatial relation to the trajectory of the manipulator, *e.g.* normal to the plane of a 2D trajectory.

Simplification of Camera Movements

In a robot-supported vision system the camera position and orientation changes dynamically for solving surveillance or inspection tasks. Different manipulator and/or camera trajectories are conceivable. As an important criterion, the trajectory should facilitate a simplification of manifolds respectively their boundaries, which is leading to efficient recognition or scoring functions. Frequently, it is advantageous to *simplify the camera movement* by decoupling rotation and translation and doing specialized movements interval by interval. For example, first the camera can be rotated such that the optical axis is directed normal to an object surface and second the camera can approach by translating along the normal direction. By log-polar transformation the manifold of transformed patterns can be represented more compactly, in which only shifts must be considered and any scaling or turning appearances are circumvented.

In a robot-supported vision system the task of object recognition can be simplified by first moving the camera in a certain relation to the object and then apply the operator, which has been learned just for the *relevant relation*. For example, by visual servoing the camera can be moved towards the object such that it appears with a certain extension in the image. The servoing procedure is based on simple, low-dimensional features like contour length or size of the object silhouette. The manifold for object recognition just represents variable object appearances which result from different view angles. However, appearance variations due to different view distances are excluded because of the prior camera movement.

Exploiting Gray Value Correlations in Time

Natural images have strong gray value correlations between neighboring pixels. A small change of the relationship between object and camera yields a small change of the image and the correlation between pixels reveal also a correlation in time.[9] For the description of the manifold respectively the boundaries between manifolds, which is the foundation for object recognition or situation scoring, we can exploit these *space-time correlations*. The use of space-time correlations will help to reduce the number of training samples and consequently reduce the effort of learning.

Extraordinary Role of Key Situations and Seed Views

The task-solving process of a camera-equipped robot system is organized as a journey in which a series of intermediate goals must be reached. Examples for such goals are intermediate effector positions in a manipulation task or intermediate camera positions in an inspection task. In the context of vision-controlled systems, the intermediate goals are key situations which are depicted in specific images (seed images, seed views). The seed views barely have gray value correlations because of large periods of time or large offsets of the points of view between taking the images. However, the seed views approximate the course of the task-solving process, and thus will serve as a framework in a servoing strategy. Later on, we will use seed views as a foundation for learning operators for object recognition and situation scoring.

---

[9] In image sequence analysis this aspect is known as *smoothness constraint* and is exploited for determining the *optical flow* [80].

3.4.2 Log-Polar Transformation for Manifold Simplification

The acquisition and application of recognition functions can be simplified by restricting the possible movements of a camera relative to a target object. Let us make the theoretical assumption (for a moment), that the possible movements just comprise a rotation around or a translation along the optical axis, and that the object is flat with the object plane normal to the optical axis. Then the rotation or scaling of appearance patterns corresponds to shifting of log-polar patterns after log-polar transformation [23]. If the camera executes the restricted movements, then a simple *cross correlation technique* would be useful to track the object in the LPT image. For this ideal case the manifold of patterns has been reduced to just a single pattern.[10]

In realistic applications, presumably, the objects are of three-dimensional shape, probably, the camera objectives cause unexpected distortions, and possibly, the optical axis is not exact normal to the object surface. Because of these realistic imponderables, certain variations of the LPT patterns occur, and the purpose of visual demonstration is to determine the actual manifold. It is expected that the manifold of LPT patterns is much more compact and easier to describe than the original manifold of appearance patterns, *e.g.* can be represented by a *single GBF with normal density distribution*.

Principles of Log-Polar Transformation

The gray value image $\mathcal{I}^G$ is separated into a foveal component around the image center which is a circle with radius ρ^{min}, and a peripheral component around the fovea which is a circle ring with maximum radius ρ^{max}. The cartesian coordinates $\{x_1, x_2\}$ of the image pixels are transformed into polar coordinates $\{\rho, \theta\}$ under the assumption of taking the center of the original image as origin. We define the log-polar transformation such that only the peripheral component is considered (shown on the left in Figure 3.20 as circle ring which is devided in sub-rings and sectors). Both the fovea and the remaining components at the image corners are suppressed (shown on the left in Figure 3.20 as gray-shaded areas). In the context of log-polar transformation it is convenient to accept values for θ in the interval $[-90°, \cdots, +270°]$. Parameter ρ takes values in the interval $[\rho^{min}, \cdots, \rho^{max}]$, with ρ^{max} the radius of the peripheral component. The set of cartesian coordinate tuples $\{(x_1, x_2)^T\}$ of the peripheral component is denoted by $\mathcal{P}^{pl}$.

Let $\mathcal{I}^L$ be the discretized and quantized *LPT image* with J_w columns and J_h rows and coordinate tuples $\{(v_1, v_2)^T\}$. For an appropriate fitting into the relevant coordinate intervals of the LPT image we define as follows.

---

[10] The proof for this theorem under the mentioned theoretical assumption is simple but will not be added, because we concentrate on realistic camera movements and pattern manifolds.

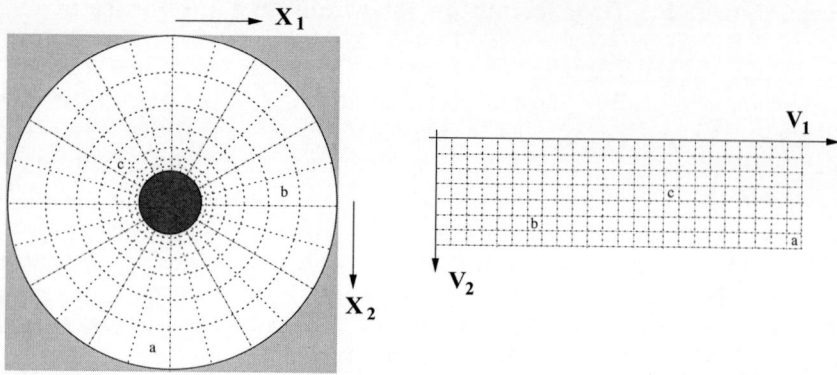

Fig. 3.20. (Left) Cartesian coordinate system $\{X_1, X_2\}$ of image pixels, partitioned into foveal circle, peripheral circle ring devided in sub-rings and sectors, and remaining components at the image corners; (Right) Log-polar coordinate system $\{V_1, V_2\}$ with horizontal axis for the angular component and the vertical axis for the radial component, mapping cake-piece sectors from (the periphery of) the gray value image into rectangle sectors of the LPT image.

$$h^a := \frac{J_w}{360°} \,, \quad h^{fo} := \log(\rho^{min}) \,,$$

$$h^{ph} := \log(\rho^{max}) \,, \quad h^b := \frac{J_h}{h^{ph} - h^{fo}} \tag{3.21}$$

Definition 3.2 (Log-polar transformation of coordinates) *The log-polar transformation of cartesian coordinates is of functionality $\mathcal{P}^{pl} \rightarrow [0, \cdots, J_w - 1] \times [0, \cdots, J_h - 1]$, and is defined by*

$$f^{v1}(x_1, x_2) := round(h^a \cdot (\theta + 90°)) \,,$$

$$f^{v2}(x_1, x_2) := round(h^b \cdot (\log(\rho) - h^{fo})) \tag{3.22}$$

$$v_1 := f^{v1}(x_1, x_2) \,, \; v_2 := f^{v2}(x_1, x_2) \tag{3.23}$$

Notice that subject to the resolution of the original image $\mathcal{I}^G$ and the resolution of the LPT image $\mathcal{I}^L$, the transformation defined by equations (3.21), (3.22), and (3.23) perhaps is not surjective, *i.e.* some log-polar pixels are undefined. However, an artificial over-sampling of the original image would solve the problem. Furthermore, we notice that in the peripheral component presumably several image pixels are transformed into just one log-polar pixel. This aspect has to be considered in the definition of log-polar transformation of gray values by taking the mean gray value from the relevant image pixels. For each log-polar pixel $(v_1, v_2)^T$ we determine the number of image pixels $(x_1, x_2)^T$ which are mapped onto this log-polar pixel.

$$f^z(v_1, v_2) := \sum_{cond} 1 , \quad with \tag{3.24}$$

$$cond := \left((x_1, x_2)^T \in \mathcal{P}^{pl}\right) \wedge \left(f^{v1}(x_1, x_2) = v_1\right) \wedge$$
$$\left(f^{v2}(x_1, x_2) = v_2\right) \tag{3.25}$$

Definition 3.3 (Log-polar transformation of gray values) *The log-polar transformation of gray values is of functionality* $[0, \cdots, J_w - 1] \times [0, \cdots, J_h - 1] \rightarrow [0, \cdots, 255]$, *and is defined by*

$$\mathcal{I}^L(v_1, v_2) := round\left(\frac{1}{f^z(v_1, v_2)} \cdot \sum_{cond} \mathcal{I}^G(x_1, x_2)\right) \tag{3.26}$$

The left and right picture in Figure 3.20 show the LPT principle, *i.e.* mapping cake-piece sectors from (the periphery of) the gray value image into rectangle sectors of the LPT image. Corresponding sectors are denoted exemplary by the symbols a,b,c. Based on the definitions of log-polar transformation, we make real-world experiments using a relevant object and an appropriate camera. The purpose of the realistic visual demonstration is to determine the actual variation of an *LPT pattern* if the object is rotating around the optical axis or the distance to the optical center of the camera is changing. Because of several imponderables in the imaging conditions and the inherent three-dimensionality of an object, we have to consider deviations from exact invariance. In the sense of the relevant discussion in Subsection 1.4.1, we are interested in the degree of compatibility between invariants of object motion and changes of the LPT patterns of the view sequence.

Experiments to Log-Polar Transformation

For this purpose an object has been put on a rotary table and the camera has been arranged such that the optical axis goes through the center of rotation with a direction normal to the rotary plane. However, this arrangement can only be reached roughly. A rotation of an appearance pattern around the image center should yield a translation of the respective LPT pattern along the axis V_1. For illustration Figure 3.21 shows two images of an *integrated circuit (IC)* object under rotation being 90° apart. These are two examples from a series 24 discrete orientations spanning equidistantly the interval $[0°, \cdots, 360°]$. Figure 3.22 shows the horizontal translation of the LPT pattern in horizontal direction, however a small deviation between the patterns occurs.

A scaling of an appearance pattern with the scaling origin at the image center should yield a translation of the LPT pattern along the axis V_2. In reality the scaling is reached by changing the view distance. Figure 3.23 shows again two images of the IC object with the same rotation angles as in Figure 3.21, but with a shorter distance to the camera. Although the object appears larger in the original images, the LPT patterns are of roughly the same size as before, but are translated in vertical direction (compare Figure 3.22 and Figure 3.24).

Fig. 3.21. Integrated circuit object under rotation by turning angle 90°, two images are taken under large viewing distance.

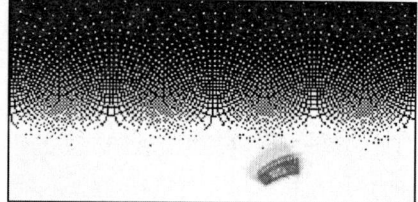

Fig. 3.22. Horizontal translation and small variation of the LPT pattern originating from the rotating object in Figure 3.21.

Variation of the LPT Patterns under Object Rotation and Scaling

An approach is needed for describing the actual variation of the LPT pattern more concretely. We present a simple technique which is based on histograms of gray values or histograms of edge orientations. First, for the set of LPT patterns the variation of the accumulations of the respective gray values are determined. This gives a measurement of possible enlargements or shrinkages of the LPT pattern. Second, for the set of LPT patterns the variation of the accumulations of the respective edge orientations are determined. This gives a measurement of possible rotations of the LPT pattern.

Our image library consists of 48 images which depict the IC object under rotation in 24 steps at two different distances to the camera (see Figure 3.21 and Figure 3.23). The histograms should be determined from the relevant area of the LPT pattern, respectively. To simplify this sub-task a nearly homogeneous background has been used such that it is easy to extract the gray value structure of the IC object.

In the first experiment, we compute for the extracted LPT patterns of the image library a histogram of gray values, respectively. Figure 3.25 (left)

Fig. 3.23. Integrated circuit object under rotation by turning angle 90°, two images are taken under small viewing distance.

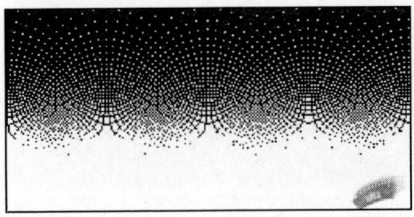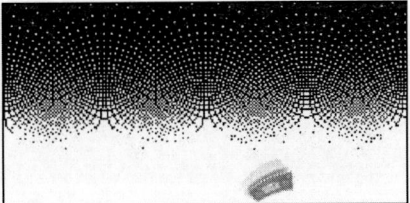

Fig. 3.24. Horizontal translation of the LPT pattern originating from the rotating object in Figure 3.21, and vertical translation compared to Figure 3.22 which is due to the changed viewing distance.

shows a histogram determined from an arbitrary image. The mean histogram is determined from the LPT patterns of the whole set of 48 images as shown in Figure 3.25 (right). Next we compute for each histogram the deviation vector from the mean histogram, which consists of deviations for each gray value, respectively. From the whole set of deviations once again a histogram is computed which is shown in Figure 3.26. This histogram resembles a Gaussian probability distribution with the maximum value at 0 and the Gaussian turning point approximately at the value ± 10. Actually, the Gaussian extent describes the difference between reality and simulation. As opposed to this, if simulated patterns are used and a perfect simulation of the imaging condition is considered, then the resulting Gaussian distribution would have the extent 0, *i.e.* the special case of an impulse.

In the second experiment, we compute for the extracted LPT patterns of the image library a histogram of gradient angles of the gray value edges, respectively. Figure 3.27 (left) shows a histogram determined from an arbitrary image. The mean histogram is determined from the LPT patterns of the whole set of 48 images as shown in Figure 3.27 (right).

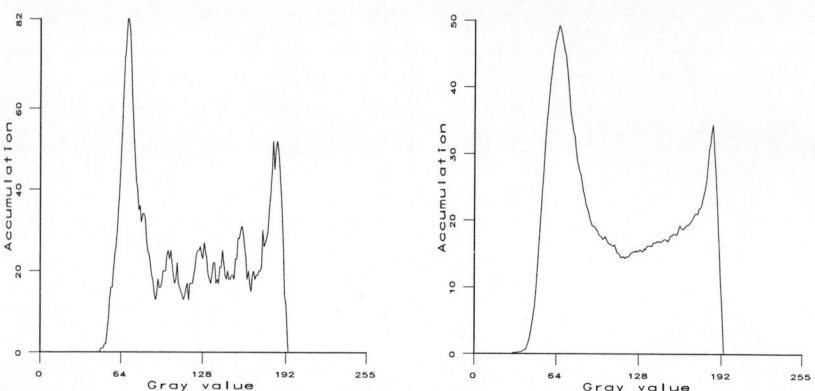

Fig. 3.25. (Left) Histogram of gray values computed from the relevant LPT object pattern of an arbitrary LPT image in Figure 3.22 or Figure 3.24; (Right) Mean histogram computed from all relevant LPT object patterns of a whole set of 48 images.

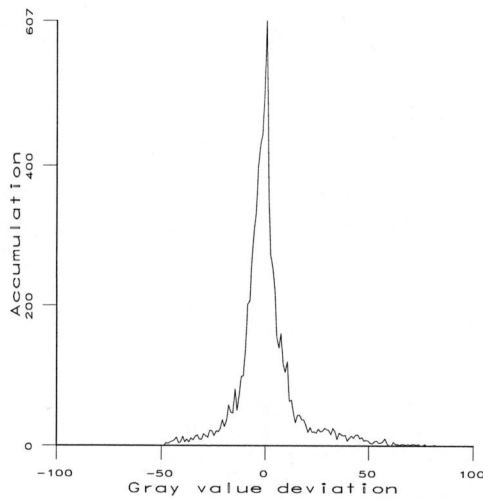

Fig. 3.26. Accumulation of gray value deviations by comparing the 48 histograms of gray values (taken from a set set of 48 images) with the mean histogram shown in Figure 3.25 (right).

Next, we compute for each histogram the deviation vector from the mean histogram. From the whole set of deviations once again a histogram is computed which is shown in Figure 3.28. This histogram can be approximated once again as a GBF with the maximum value at 0 and the Gaussian turn-

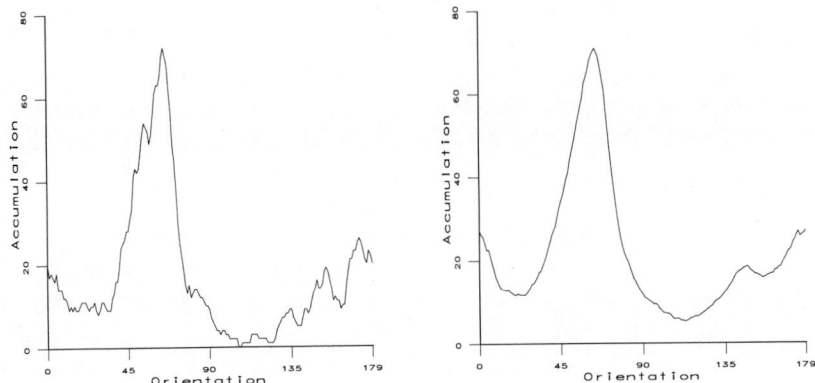

Fig. 3.27. (Left) Histogram of edge orientations computed from the relevant LPT object pattern of an arbitrary LPT image in Figure 3.22 or Figure 3.24; (Right) Mean histogram computed from all relevant LPT object patterns of a whole set of 48 images.

ing point approximately at the value ±5. In a simulated world the Gaussian distribution would have the extent 0.

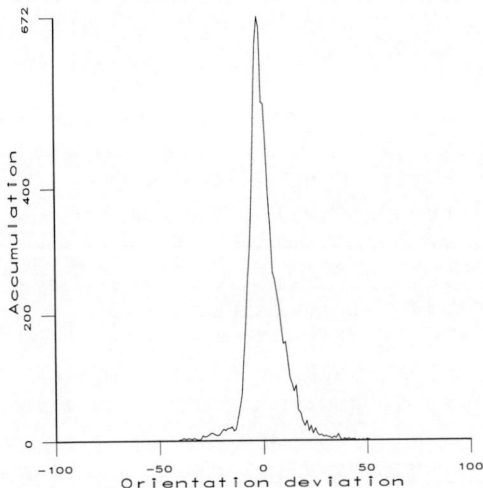

Fig. 3.28. Accumulation of orientation deviations by comparing the 48 histograms of edge orientation (taken from a set of 48 images) with the mean histogram shown in Figure 3.27 (right).

Compatibility Properties under Log-Polar Representation

These experiments show exemplary that in reality the log-polar patterns change slightly when the object is rotating or moving towards the camera. The theoretical concept of invariance must be relaxed by the practical concept of compatibility. Related to the Gaussian approximation of the histograms, the invariance is characterized by the special GBF with extent 0, but compatibility is characterized by an extent beyond 0.

The compatibility properties under log-polar representation are useful most of all for robot-supported vision systems with active cameras. For detailed object inspection a camera should approach in normal direction to the object plane, because with this strategy the tracking procedure of LPT patterns is simplified. In this case, the relevant manifold of LPT patterns, which is represented by the recognition function, simply consists of one pattern together with slight deviations. Actually, the degree of compatibility determines the acceptance level of the recognition function. If the recognition function is represented by a GBF network (see Subsection 3.2.2) then one single GBF is supposed to be enough, whose center vector is specified by the typical LPT pattern and whose Gaussian extent is specified by the accepted deviations.

Log-Polar Representation of Images of 3D Objects

However, if objects are of three-dimensional shapes and/or are sequentially viewed under non-normalized conditions, then LPT will not reduce the complexity of pattern manifolds. Obviously, the compatibility properties under log-polar representation are invalid if the observed object is of three-dimensional shape, because the normal directions of certain faces differ significantly from the optical axis of the camera and certain object faces appear or disappear under different viewing angles. For tall objects both the top face and side faces come into play, but all of them can not be orthogonal to the direction of camera movement. However, a conceivable approach is to work only with the top face of an object. Perhaps, the boundary of the top face can be extracted with the approaches in Chapter 2.

The nice properties of LPT are likewise invalid if the rotation axis and the optical axis of the camera differ significantly. It is the purpose of visual demonstration and experimentation to determine for certain camera movements the degrees of compatibility and find out a movement strategy such that a certain level of simplicity of the tracking process is reached. The measured degree of compatibility can be used as criterion for arranging the camera prior to the application phase automatically. For example, in a servoing approach the camera can be rotated such that the optical axis is directed normal to the object plane (see Chapter 4). In the application phase the camera can move along the optical axis and a simple cross-correlation technique would be useful to track the object in the LPT images. This brief discussion drew attention to the following important aspect.

For simplifying the appearance manifold, the log-polar transformation must be combined with procedures for boundary extraction and must also rely on servoing techniques for reaching and keeping appropriate camera-object relations.

Depending on the specific task, performed in the relevant environment, the conditions for applying log-polar transformation may or may not be attainable. The following subsection assumes that LPT can not be applied. Instead, an approach for a sophisticated approximation of manifolds is presented which exploits space-time correlations in an active vision application.

3.4.3 Space-Time Correlations for Manifold Refinement

The approach assumes that the pattern variation of the manifold can be represented as a one-dimensional course, approximately. In an active vision application, the assumption holds if a camera performs a one-dimensional trajectory around a stationary object, or the object moves in front of a stationary camera. More complicated pattern variations, probably induced by simultaneous movements of both camera and object or additional changes of lighting conditions, are not accepted. Restricted to the mentioned assumption, we present a strategy for refined manifold approximation which is based on a GBF network with a specific category of hyper-ellipsoidal basis functions. The basis functions are stretched along one direction, which is determined on the basis of the one-dimensional space-time correlations.

Variation of the Gray Value Patterns under Object Rotation

As opposed to the previous subsection, in the following experiments we do not apply LPT. Instead, we make measurements in the original gray value images, but restrict our attention to the distribution of edge orientations. A three-dimensional transceiver box is put on a rotary table which is rotating in discrete steps with offset $5°$, *i.e.* altogether 72 steps in the interval $[0°, \cdots, 360°]$. For each step of rotation a camera takes an image under a constant non-normal viewing direction of angle $45°$ relative to the table. Figure 3.29 shows four images from the whole collection of 72. The computation of gradient magnitudes followed by a thresholding procedure yields a set of gray value edges, as shown in the binary image, respectively.

Related to each binary image we compute from the set of edges the discretized orientations, respectively, and determine the histogram of edge orientations. The discretization is in steps of $1°$ for the interval $[0°, \cdots, 180°]$. Figure 3.30 shows in the left diagram an overlay of four histograms from the four example images depicted previously, and shows in the right diagram the mean histogram computed from the whole collection of 72 images.

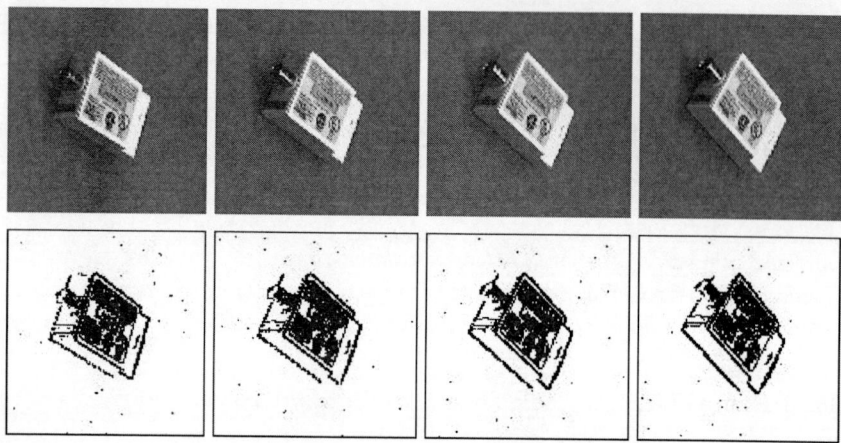

Fig. 3.29. (Top) Four gray value images of a transceiver box under rotation in discrete steps of turning angle 5°; (Bottom) Binarized images of extracted gray value edges.

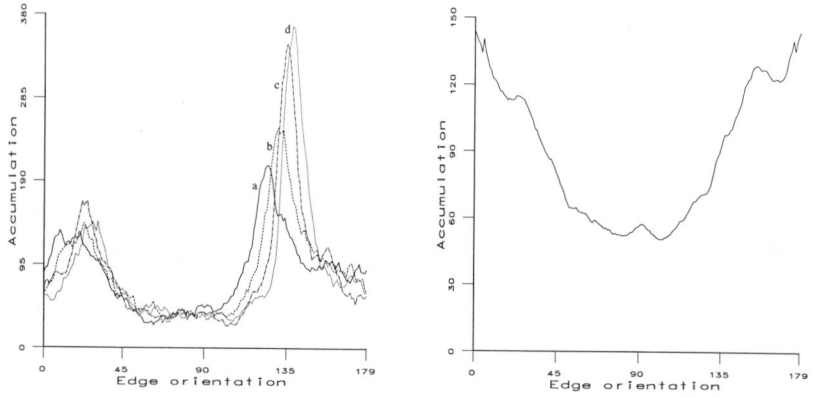

Fig. 3.30. (Left) Overlay of four histograms of edge orientations computed for the four images in Figure 3.29, (Right); Mean histogram of edge orientations computed from a whole set of 72 images.

The distribution of deviations from the mean histogram is shown in Figure 3.31. The main difference when comparing this distribution with that in Figure 3.28 is that the maximum accumulation is not reached for the value 0° of orientation deviation but is reached far beyond (approximately value −70°), *i.e.* the compatibility property under rotation does not hold.

For general situations like these, the manifold of appearance patterns can be represented by a GBF network consisting of more than one GBF (see Subsection 3.2.2). We are interested in a sparse approximation of the recognition function which is characterized by a minimum number of GBFs.

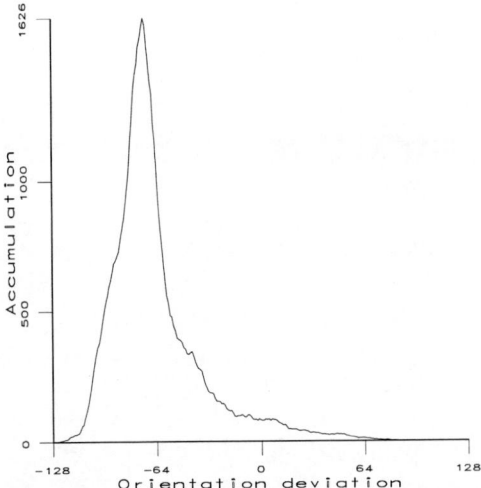

Fig. 3.31. Accumulation of orientation deviations by comparing the 72 histograms of edge orientation, taken from a set of 72 images, with the mean histogram shown in Figure 3.30 (right).

In Section 3.3, a reduction of the number of GBFs has been reached with a clustering procedure which approximates each sub-set of similar patterns by just one typical pattern, respectively. However, we completely disregarded that the application of recognition functions takes place in a task-solving process, in which the relation between object and camera changes continually. The *principle of animated vision* should also be considered in the learning procedure, which will help to reduce the effort of clustering. In consensus with a work of Becker [16], one can take advantage of the *temporal continuity* in image sequences.

Temporal Continuity between Consecutive Images

The temporal continuity can be observed exemplary in a series of histograms of edge orientations for an object under rotation. Figure 3.30 (left) showed four histograms (a,b,c,d) for the object in Figure 3.29, which has been rotated slightly in four discrete steps of 5°, respectively. The histogram curves moved to the right continually under slight object rotation.[11]

   A further example of temporal continuity is based on gray value correlations within and between images. For illustration, once again the transceiver box is used, but now images are taken in discrete steps of 1° in the orientation interval $[0°, \cdots, 35°]$. Figure 3.32 shows eight example images from the collection of 36 under equidistant angle offsets. The correlations can be observed

---

[11] The variation of the accumulation values is due to changing lighting conditions or due to the appearing or disappearing of object faces.

easily by extracting the course of gray values at a single image pixel. For example, we selected two such pixels which have been marked in the images of Figure 3.32 by a black and a white dot, respectively.

Fig. 3.32. Eight example images from the collection of 36 under equidistant turning angles, and overlay with a white and a black dot at certain pixel positions.

The courses of gray values at these two points are shown in the diagrams of Figure 3.33. For certain intervals of time a piece-wise linear approximation of the gray value variation seems to be appropriate. This piece-wise linearity is an indication for reasonably assuming space-time correlation of gray values.

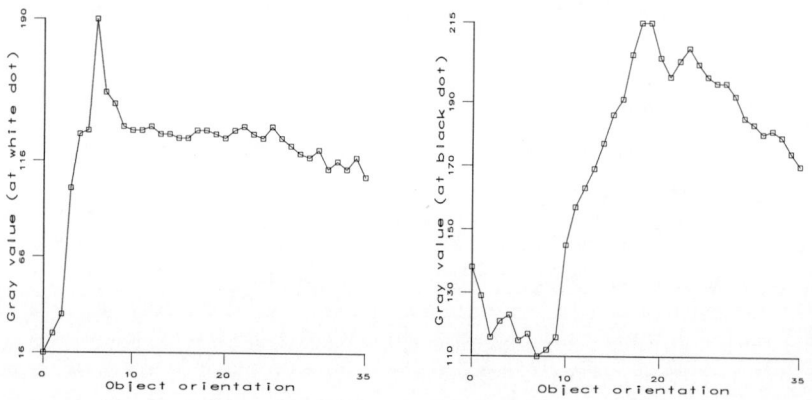

Fig. 3.33. (Left) Course of gray value at position of white dot in Figure 3.32 for 36 images of the rotating object; (Right) Relevant course at position of black dot.

Generally, gray value correlations only hold for small spatial distances in the image or short time durations in a dynamic object/camera relationship. This aspect must be considered when exploiting the space-time correlations for the manifold approximation. We would like to approximate the manifold on the basis of a sparse set of training data, because this will reduce the effort of learning. Actually, this requirement of sparseness can be supported by making use of the space-time correlations.

Considering Space-Time Correlations at Seed Views

We determine space-time correlations for a small set of seed views and construct specific GBFs thereof, so-called *hyper-ellipsoidal Gaussian basis functions*. As already discussed at the beginning of this Subsection 3.4.3, we assume a one-dimensional course of the manifold approximation. Therefore, each GBF is almost hyper-spherical except for one direction whose GBF extent is stretched. The exceptional direction at the current seed view is determined on the basis of the difference vector between the previous and the next seed view. Actually, this approach incorporates the presumption of approximate, one-dimensional space-time correlations.

For illustrating the principle, we take two-dimensional points which represent the seed views. Figure 3.34 shows a series of three seed views, *i.e.* previous, current and next seed view. At the current seed view the construction of an elongated GBF is depicted. Actually, an ellipse is shown which represents the contour related to a certain Gaussian altitude.

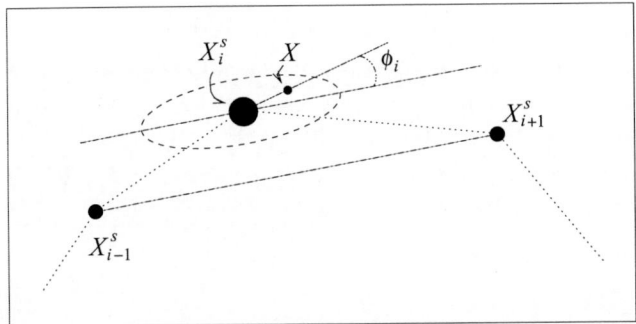

Fig. 3.34. Principle of constructing hyper-ellipsoidal basis functions for time-series of seed vectors.

The GBF extent along this exceptional direction must be defined such that the significant variations between successive seed views are considered. For orthogonal directions the GBF extents are only responsible for taking random imponderables into account such as lighting variations. Consequently, the GBF extent along the exceptional direction must be set larger than the extent along the orthogonal directions. It is reasonable to determine the exceptional

GBF extent dependent on the euclidean distance measurement between the previous and the next seed view.

Generally, the mentioned approach can be formalized as follows. Let $(X_1^s, \cdots, X_I^s)$ be the time series of seed vectors. A hyper-spherical Gaussian f_i^{Gs} is defined with the center vector X_i^s, and extent parameter σ_i is set equal to 1. Later on, it will become obvious that there is no loss of generality involved with the constant extent value.

$$f_i^{Gs}(X) := exp\left(-\|X - X_i^s\|^2\right) \tag{3.27}$$

The Gaussian computes equal values for vectors X located on an m-dimensional hyper-sphere around the center vector X_i^s. However, we are interested in an m-dimensional hyper-ellipsoid. Specifically, $(m-1)$ ellipsoid axes should be of equal half-lengths, i.e. $\kappa_2 = \kappa_3 = \cdots = \kappa_m$, and one ellipsoid axis with half-length κ_1, which should be larger than the others. For this purpose the hyper-spherical Gaussian is modified as follows. We take in the time-series of seed vectors relative to the current vector X_i^s the previous vector X_{i-1}^s and the next vector X_{i+1}^s. Let us define two difference vectors

$$\Delta_i^1 := X - X_i^s \ , \quad \Delta_i^2 := X_{i+1}^s - X_{i-1}^s \tag{3.28}$$

The angle ϕ_i between both difference vectors is computed as

$$\phi_i := \arccos\left(\frac{\left(\Delta_i^1\right)^T \cdot \Delta_i^2}{\|\Delta_i^1\| \cdot \|\Delta_i^2\|}\right) \tag{3.29}$$

The modifying expression for the Gaussian is defined as

$$f_i^{mg}(X) := \sqrt{(\kappa_1 \cdot \cos(\phi_i))^2 + (\kappa_2 \cdot \sin(\phi_i))^2} \tag{3.30}$$

Parameter κ_1 is defined on the basis of the euclidean distance between vectors X_{i-1}^s and X_{i+1}^s, e.g. as the half of this distance. Parameter κ_2 may be defined by a certain percentage α of κ_1, $\alpha \in [0, \cdots, 1]$.

$$\kappa_1 := \frac{\|\Delta_i^2\|}{2} \ , \quad \kappa_2 := \alpha \cdot \kappa_1 \tag{3.31}$$

Finally, the modified Gaussian is as follows

$$f_i^{Gm}(X) := f_i^{Gs}(X) \cdot f_i^{mg}(X) \tag{3.32}$$

It can be proven, that equation (3.32) is equal to a Gaussian with Mahalanobis distance between input vector X and center vector X_i^s. The underlying hyper-ellipsoid is of the specific form as explained above.

Although the approach is very simple, both efficiency and robustness of the recognition function increases significantly (see later on).

Validation of Space-Time Correlations at Seed Views

A realistic motivation of the principle can be presented for the series of 36 images of the transceiver box, e.g. a subset of eight has already been shown in Figure 3.32. We define symbols $\mathcal{I}_l^G \in \{\mathcal{I}_0^G, \cdots, \mathcal{I}_{35}^G\}$ to designate the images. For each image $\mathcal{I}_l^G$, we take the two gray values at the positions marked by the white and black dot and combine them to a two-dimensional feature vector X_l, respectively, *i.e.* altogether 36 possible vectors. The two courses of gray values have already been shown in Figure 3.33. A subset of eight seed images $\mathcal{I}_0^G, \mathcal{I}_5^G, \cdots, \mathcal{I}_{35}^G$ is selected with equidistant angle offset 5°. Actually these are the ones depicted in Figure 3.32. Based on the feature vector in each seed image, we obtain the seed vectors $X_1^s := X_0, X_2^s := X_5, \cdots, X_8^s := X_{35}$. By applying the approach summarized in equation (3.32), it is possible to construct an elongated Gaussian around the seed vectors $X_2^s, \cdots, X_7^s$, respectively. In Figure 3.35, each picture represents the two-dimensional feature space for the pair of gray values. The big dot marks the seed vector X_i^s and the two dots of medium size mark the seed vectors X_{i-1}^s and X_{i+1}^s, respectively. The small dots represent the feature vectors X_l collected from a series of 11 images $\mathcal{I}_l^G$, half by half taken prior and after the image which belongs to the seed vector X_i^s, *i.e.* these are the images $\mathcal{I}_l^G \in \{\mathcal{I}_k^G, \cdots, \mathcal{I}_n^G\}$, with $k := (i-2) \cdot 5, n := i \cdot 5$. We observe for nearly all pictures in Figure 3.35, that the constructed ellipses approximate the distribution of relevant feature vectors quite good. Especially, the ellipsoid approximation is more appropriate than circles which would originate from radial symmetric Gaussians.

Exploiting Space-Time Correlations for Object Recognition

The usefulness of constructing elongated Gaussians can be illustrated for the task of object recognition. As opposed to the previous example in which a two-dimensional feature space of pairs of gray values has been used, in the following we will consider histograms of edge orientations. The left diagram in Figure 3.30 shows four example histograms which have been determined from images of the transceiver box under four successive rotation steps. The angles are discretized in integer values of the set $\{0, \cdots, 179\}$ and therefore the feature vector consists of 180 components.

We would like to construct a recognition function which makes use of the temporal continuity involved in object rotation. For the purpose of training, the transceiver box is rotated in steps of 10° and all 36 training images are used as seed images. The computation of gradient magnitudes followed by a thresholding procedure yields a set of gray value edges for each seed image, respectively. From each thresholded seed image a histogram of edge orientations can be computed. A GBF network is learned by defining elongated GBFs according to the approach presented above, *i.e.* using the histograms of the seed images as the Gaussian center vectors and modifying the Gaussians based on previous and next seed histograms. Parameter κ_1 in equation (3.30) is specified as half of the euclidean distance between the previous and the next seed vector (X_{i-1}^s and X_{i+1}^s), and κ_2 is specified by $0.33 \cdot \kappa_1$. In the GBF

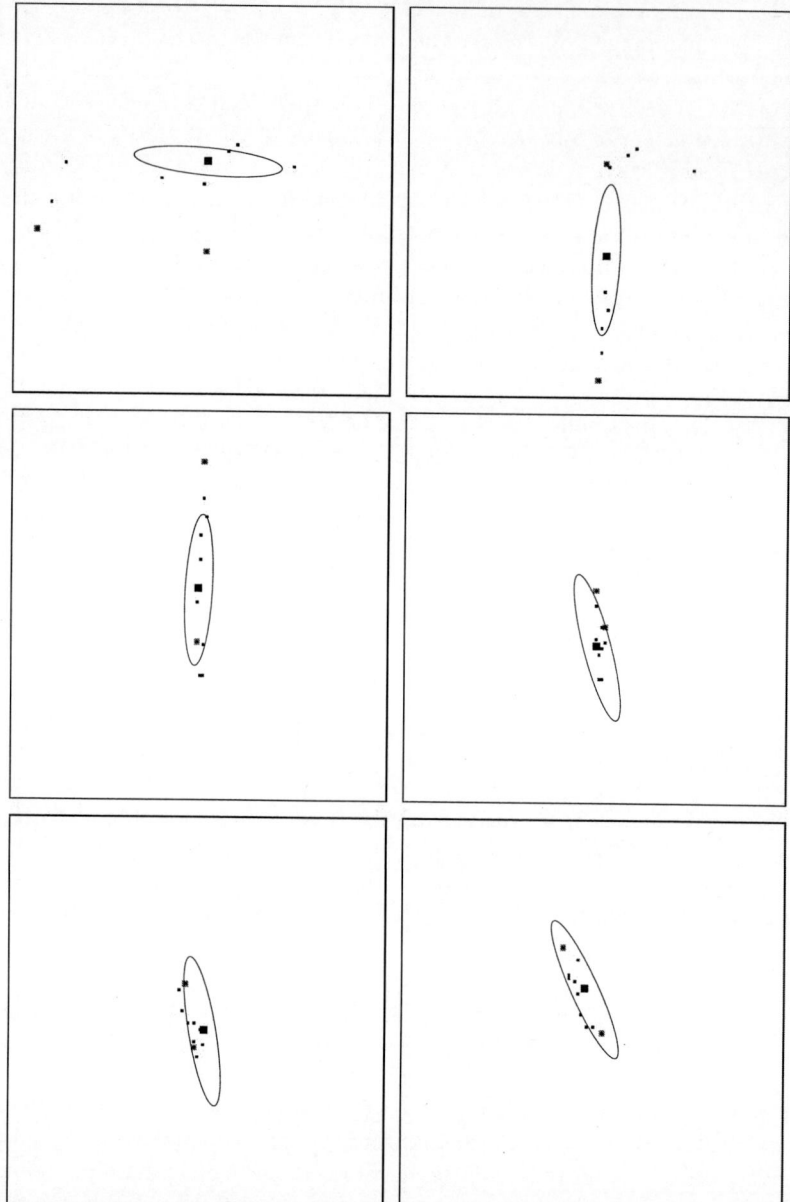

Fig. 3.35. Six pictures showing the construction of elongated Gaussians for six seed vectors (originating from six seed images), each picture represents the two-dimensional feature space of a pair of gray values, the black dots represent the gray value pair belonging respectively to images prior and after the seed image.

network the combination factors for the Gaussians are determined by the pseudo inverse technique.[12]

For assessing the network of elongated Gaussians, we also construct a network of spherical Gaussians and compare the recognition results computed by the two GBF networks. The testing views are taken from the transceiver box but different from the training images. Actually, the testing data are subdivided in two categories. The first category consists of histograms of edge orientations arising from images with a certain angle offset relative to the training images. Temporal continuity of object rotation is considered purely. For these situations the relevant recognition function has been trained particularly. The second category consists of histograms of edge orientations arising from images with angle offset and are *scaled*, additionally. The recognition function composed of elongated Gaussians should recognize histograms of the first category robustly, and should discriminate clearly the histograms of the second category. The recognition function composed of spherical Gaussians should not be able to discriminate between both categories, which is due to an increased generalization effect, *i.e.* accepting not only the angle offsets but also scaling effects.

The desired results are shown in the diagrams of Figure 3.36. By applying the recognition function of spherical Gaussians to all testing histograms, we can hardly discriminate between the two categories. Instead, by applying the recognition function of elongated Gaussians to all testing histograms, we can define a threshold for discriminating between both categories.

Discussion of the Learning Strategy

Based on the results of these experiments, we draw the following conclusion concerning the strategy of learning. For a continually changing relation between object and camera we have observed space-time correlations of gray values. That is, the orbit of patterns is piece-wise continuous and can be approximated piece-wise linear. In consequence of this, only a reduced set of seed views needs to be taken into account for learning a recognition function. From the seed views we not only use the individual image contents but also the gray value relations between successive seed views.

> In the process of learning operators for object recognition the temporal continuity of training views can be exploited for the purpose of reducing the number of views.

This motivates the use of a sparse set of elongated Gaussians for constructing a fine-tuned approximation of recognition functions. The approach overcomes the need of a large set of training views and avoids the time-consuming

---

[12] A simple post-processing strategy is conceivable for reducing the number of GBFs. According to this, neighboring seed vectors X_i^s and X_{i+1}^s can be collected if the respective difference vectors Δ_i^2 and Δ_{i+1}^2 are approximately collinear.

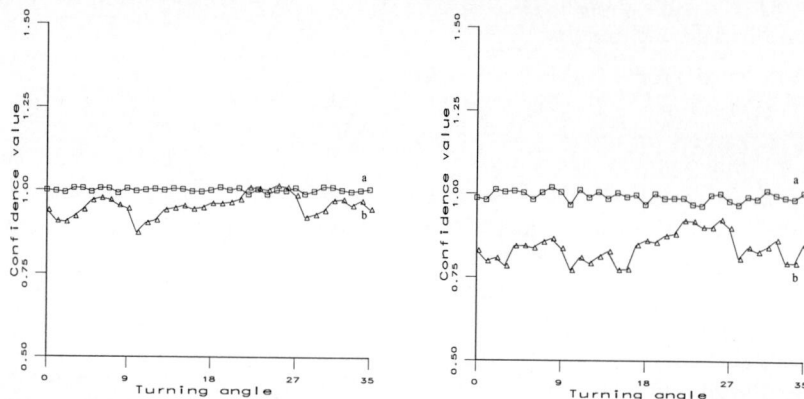

Fig. 3.36. Confidence values of recognizing an object based on histograms of edge orientations. For testing, the object has been rotated by an offset angle relative to the training images (result given in curve a), or the object has been rotated and the image has been scaled additionally relative to the training images (result given in curve b). (Left) Curves show the courses under the use of spherical Gaussians, both categories of testing data can hardly be distinguished; (Right) Curves show the courses under the use of elongated Gaussians, both categories of testing data can be distinguished clearly.

clustering process. In our previous experiments, we took the training views in discrete steps by changing a certain degree-of-freedom, *e.g.* constant angle offset of a rotating object, and defined this set as seed views. Strategies are needed for determining the appropriate offset automatically. Alternatively, more sophisticated approaches are conceivable for choosing a relevant set of seed views, as discussed in Section 3.5.

3.4.4 Learning Strategy with PCA/GBF Mixtures

In Subsection 3.2.1, the learning of a recognition function has been embedded into the problem of estimating the parameters of an implicit function. The set of example vectors which fulfills the function equation approximates the pattern manifold of the relevant object. For realistic applications a certain degree of inequality must be accepted, *i.e.* pure invariance is relaxed to compatibility. In Subsection 3.2.2, it was suggested to represent the implicit function by a network of Gaussian basis functions, because universal approximations can be reached by varying the number of GBFs. For fine-tuning the GBFs one should exploit temporal continuity in acquiring the training views, as was motivated in Subsection 3.4.3. In Subsection 3.2.3, the implicit function has been defined alternatively as a hyper-ellipsoid, whose principal axes and half-lengths are determined by principal component analysis of a small set of seed views. The number of principal axes is equal to the number of seed views and this number is much less than the size of an object pattern.

In consequence of this, the principal axes specify a canonical frame whose dimension is much less than the dimension of the input space.

The Role of Counter Situations in Learning

Both the GBF and the PCA approach to learning do not consider counter situations directly for acquiring the recognition functions. Instead, for a set of typical patterns of an object, a function is learned which responds almost invariant. The degree of compatibility is used as threshold for discriminating counter situations. In certain robot tasks this learning strategy is the only conceivable one, *e.g.* if a training set of views is available just for one specific object which should be detected in any arbitrary environment. However, in many robot applications it is realistic that certain environments or certain situations occur more frequent than others. It is important to consider counter situations from typical environments for fine-tuning a recognition function and thus increasing the robustness of recognition.

Survey to the Coarse-to-Fine Strategy of Learning

We present a *coarse-to-fine strategy* of learning a recognition function, which approximates the manifold of object patterns coarsely from a sparse set of *seed views* and fine-tunes the manifold with more specific object patterns or counter situations, so-called *validation views*. For the coarse approximation of the manifold either the GBF or the PCA approach is suitable. For fine-tuning the manifold we use once again a GBF network. Both the coarse and the fine approximation are controlled under certain PAC requirements (see Definition 3.1). The function for object recognition should be PAC-learned subject to the probability P^r and threshold ζ. It is reasonable to choose for parameter P^r a positive real value near the maximum 1 (value 1 means 100 percent) and also for parameter ζ a positive value near 1.

   Let f^{Gi} be an implicit function according to equation (3.5), with parameter vector B and input-output vector Z. The vector Z is composed of the input component X, representing a pattern, and the output component Y, representing a class label or scoring value. In this subsection we focus on the classification problem. We will have several functions f_k^{Gi} each responsible for a certain class with label k. In the following, label k is suppressed to avoid the overload of indices. For convenience, we also suppress label k in vector Z and instead accept vector X in the application of function $f^{Gi}(B, X)$.

Coarse Approximation Based on Seed Patterns

The first step, *i.e.* coarse approximation, is based on an *ensemble of seed patterns* $X_i^s \in \mathcal{X}^s$. It is assumed that function f^{Gi} has been PAC-learned from the seed views subject to parameters P^r and ζ with either the GBF or the PCA approach. The PAC requirement holds trivially with $P^r = 1$ and $\zeta = 1$, because both approaches are configured such that all seed patterns are located on the orbit, respectively. Related to Definition 3.1 of PAC-recognition, in this first step the target patterns are the seed patterns and counter patterns are not considered.

Fine Approximation Based on Validation Patterns

In the second step, *i.e.* fine approximation, we take an *ensemble of validation patterns* X_j^v into account which is subdivided into two classes. The first class $\mathcal{X}^{vp}$ of validation patterns is extracted from additional views of the target object and the second class $\mathcal{X}^{vn}$ of patterns is extracted from views taken from counter situations. Depending on certain results of applying the implicit function f^{Gi} (computed in the first step) to these validation patterns we specify spherical Gaussians according to equation (3.10) and combine them appropriately with the definition of the implicit function. The purpose is to modify the original implicit function and thus fine-tune the approximation of the pattern orbit of the target object, *i.e.* target patterns should be included and counter patterns excluded.

For each validation pattern $X_j^v \in \mathcal{X}^{vp} \cup \mathcal{X}^{vn}$ we apply function f^{Gi} which yields a *measurement of proximity* to the coarsely learned pattern orbit.

$$\eta_j := f^{Gi}(B, X_j^v) \tag{3.33}$$

For $\eta_j = 0$ the pattern X_j^v is far away from the orbit, for $\eta_j = 1$ the pattern belongs to the orbit. There are *two cases* for which it is reasonable to modify the implicit function. *First*, maybe a pattern of the target object is too far away from the orbit, *i.e.* $X_j^v \in \mathcal{X}^{vp}$ and $\eta_j \leq \zeta$. *Second*, maybe a pattern of a counter situation is too close to the orbit, *i.e.* $X_j^v \in \mathcal{X}^{vn}$ and $\eta_j \geq \zeta$. Pattern X_j^v is the triggering element for fine-tuning the coarse approximation. In the first case, the modified function should yield value 1 at pattern X_j^v, and in the second case, should yield value 0. Additionally, we would like to reach generalization effects in the local neighborhood of this pattern. A prerequisite is to have continuous function values in this local neighoorhood including the triggering pattern X_j^v.

The modification of the implicit function takes place by locally putting a spherical Gaussian f_j^{Gs} into the space of patterns, then multiplying a weighting factor to the Gaussian, and finally adding the weighted Gaussian to the implicit function. The mentioned requirements can be reached with the following parameterizations. The center vector of the Gaussian is defined by the relevant pattern X_j^v.

$$f_j^{Gs}(X) := exp\left(-\frac{1}{\tau} \cdot \|X - X_j^v\|\right) \tag{3.34}$$

For the two cases, we define the weighting factor w_j for the Gaussian individually. It will depend on the computed measurement of proximity η_j of the pattern X_j^v to the coarsely learned pattern orbit.

$$w_j := \begin{cases} 1 - \eta_j & : \quad target\ pattern\ too\ far\ away\ from\ orbit \\ -\eta_j & : \quad counter\ pattern\ too\ close\ to\ orbit \end{cases} \tag{3.35}$$

The additive combination between the implicit function and the weighted Gaussian yields a new function for which the orbit of patterns has been changed. By considering the above definitions in equation (3.33), (3.34), and (3.35, the modified function will meet the requirements. In particular, the desired results are obtained for the triggering pattern, *i.e.* $X = X_j^v$. In the first case, we assume a weighted Gaussian has been constructed for a vector $X_j^v \in \mathcal{X}^{vp}$. By applying the vector to the modified function the value 1 is obtained, *i.e.* the pattern belongs to the orbit. In the second case, we assume that a weighted Gaussian has been constructed for a vector $X_j^v \in \mathcal{X}^{vn}$. By applying the vector to the modified function the value 0 is obtained, *i.e.* the pattern is far away from the orbit. In both cases, the Gaussian value is 1, and the specific weight plays the role of an increment respective decrement to obtain the final outcome 1 respective 0.

In the neighborhood (in the pattern space) of the triggering pattern the modified function produces a smoothing effect by the inherent extension of the Gaussian. It is controlled via factor τ in equation (3.34). The higher this factor the larger the local neighborhood of X_j^v which is considered in modifying the manifold. The size of this neighborhood directly corresponds with the degree of generalization.

Constructing the Recognition Function

With this foundation we show how to construct recognition functions. Generally, the ensemble of validation patterns contains several triggering patterns. We determine the set of triggering patterns and define for each one a Gaussian, respectively. Let $j \in 1, \cdots, J$ be the indices of the set of necessary Gaussians as explained above. The overall recognition function f^{rc} can be defined as the sum of the transformed implicit function and the linear combination of necessary Gaussians.

$$f^{rc}(X) := f^{Gi}(B, X) + \sum_{j=1}^{J} w_j \cdot f_j^{Gs}(X) \tag{3.36}$$

Vector X represents an unknown pattern which should be recognized. The parameter vector B has been determined during the phase of coarse approximation. The number and the centers of Gaussians are constructed during the phase of fine approximation. There is only one degree of freedom which must be determined, *i.e.* factor τ for specifying the extent of the Gaussians. Iterative approaches such as the *Levenberg-Marquardt algorithm* can be applied for solution [134, pp. 683-688].

Criterion for the Recognition of Unknown Patterns

Based on a coarse approximation of the pattern manifold of the target object (using seed views), the approach has taken a further set of patterns into account arising from the target object and from counter situations (using validation views), in order to fine-tune the approximation and make the recognition function more robust. This coarse-to-fine strategy of learning can be applied to any target object which we would like to recognize. If $k \in \{1, \cdots, K\}$ is the index for a set of target objects, then recognition functions f_k^{rc}, with $k \in \{1, \cdots, K\}$, can be learned as above. For a robust discrimination between these target objects it is reasonable to learn recognition functions for target objects by considering the other target objects as counter situations, respectively. The final decision for classifying an unknown pattern X is by looking for the maximum result computed from the set of recognition functions f_k^{rc}, $k \in \{1, \cdots, K\}$.

$$k^* := \arg \max_{k \in \{1, \cdots, K\}} f_k^{rc}(X) \tag{3.37}$$

Aspects of PAC Requirements to the Recognition Function

All recognition functions f_k^{rc} must be PAC-learned subject to the parameters P^r and ζ. The PAC requirement must be checked for the whole set of training data, which consist of the seed patterns and the additional validation patterns of the target object, and the patterns of counter situations. If this requirement does not hold for certain recognition functions, it is necessary to increase the number of seed vectors and thus increase the dimension of the pattern space in which the fine approximation takes place. In consequence of this, each recognition function approximates a manifold of patterns, whose dimension depends on the difference or similarity compared to other target objects in the task-relevant environment. The recognition function for an object with easy discrimination versus other objects is defined in a low-dimensional space, and in case of hard discrimination the function is defined in a high-dimensional space. This strategy of learning recognition functions is in consensus with the design principle of purposiveness (see Section 1.2), *i.e.* subject to the requirements of the task the recognition function is constructed with *minimum description length*.

Visualization of the Approach Applied to Pattern Recognition

The coarse-to-fine strategy of learning can be illustrated graphically. Similar to Section 3.2 we assume three seed patterns X_1^s, X_2^s, X_3^s which are treated as points in the high-dimensional input space. By using principal component analysis for coarse approximation, we obtain an ellipse through the points. The exponential transformation of the implicit function yields values of proximity to the ellipse between 0 and 1. Actually, the course of proximity values obtained from a straight course of points passing the orbit perpendicular is a

Gaussian (see Figure 3.37). Let us assume that the extent of this Gaussian is equal to the extent of Gaussians which are taken into account for fine-tuning the coarse approximation.

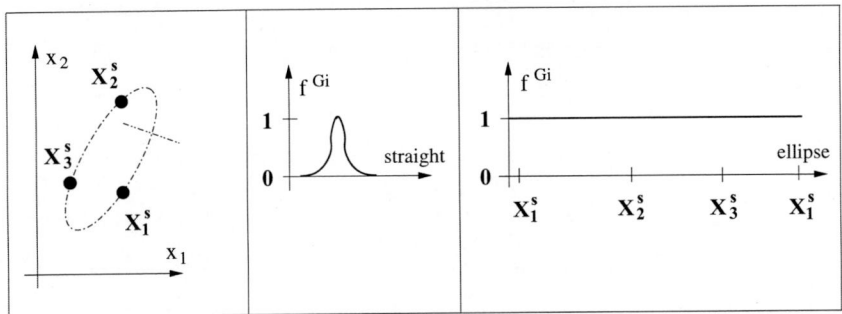

Fig. 3.37. (Left) Ellipse through three seed vectors and perpendicular straight line across the ellipse; (Middle) Gaussian course of proximity values along the straight line, (Right) Constant course of proximity values along the ellipse.

As a first example, it may happen that a pattern of a counter situation, *i.e.* $X_1^v \in \mathcal{X}^{vn}$, is located on the ellipse (see Figure 3.38 (left)). A Gaussian f_1^{Gs} is defined with X_1^v as center vector, and weight $w_1 := 1$. The combination of implicit function and weighted Gaussian according to equation (3.36) decreases the value of recognition function f^{rc} locally around point X_1^v. Figure 3.38 (middle) shows the effect along the straight course of points passing the orbit perpendicular. A positive and a negative Gaussian are added which yields constant 0. Figure 3.38 (right) shows the values of the recognition function f^{rc} along the course of the ellipse, which are constant 1 except for the Gaussian decrease to 0 at point X_1^v.

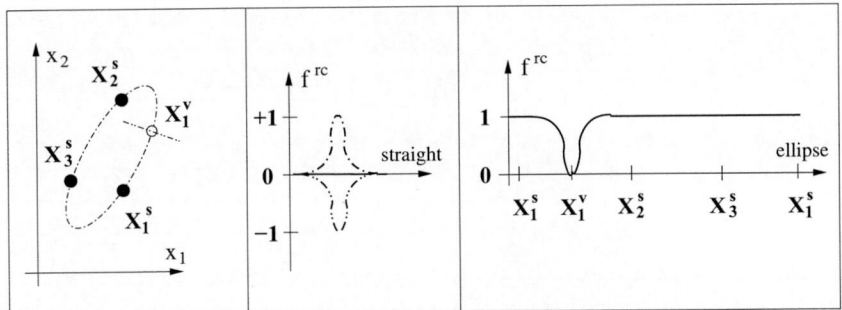

Fig. 3.38. (Left) Ellipse through three seed vectors and perpendicular straight line through a counter vector located on the ellipse; (Middle) Along the straight line the positive Gaussian course of proximity values is added with the negative Gaussian originating from the counter vector, resulting in 0; (Right) Along the ellipse the recognition values locally decrease at the position of the counter vector.

The second example considers a further counter pattern, *i.e.* $X_2^v \in \mathcal{X}^{vn}$, which is too near to the ellipse orbit but not located on it. Figure 3.39 shows similar results compared to the previous one, however the values of function f^{rc} along the course of the ellipse are less affected by the local Gaussian.

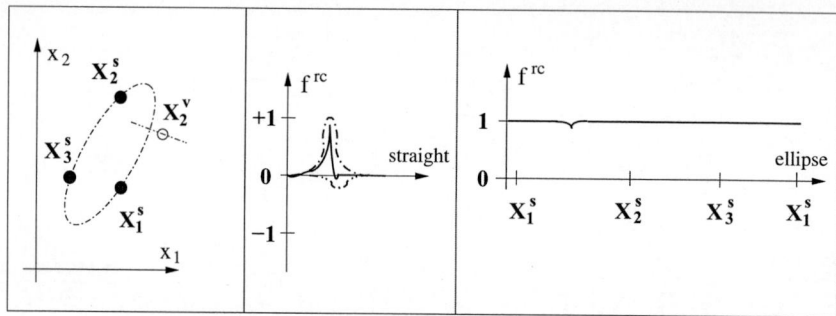

Fig. 3.39. (Left) Ellipse through three seed vectors and perpendicular straight line through a counter vector located near to the ellipse; (Middle) Along the straight line the positive Gaussian course of proximity values is added with the shifted negative Gaussian originating from the counter vector, such that the result varies slightly around 0; (Right) Along the ellipse the recognition values locally decrease at the position near the counter vector, but less compared to Figure 3.38.

The third example considers an additional pattern from the target object, *i.e.* $X_3^v \in \mathcal{X}^{vp}$, which is far off the ellipse orbit. The application of f^{Gi} at X_3^v yields η_3. A Gaussian is defined with vector X_3^v taken as center vector, and the weighting factor w_3 is defined by $(1 - \eta_3)$. The recognition function is constant 1 along the course of the ellipse, and additionally the function values around X_3^v are increased according to a Gaussian shape (see Figure 3.40).

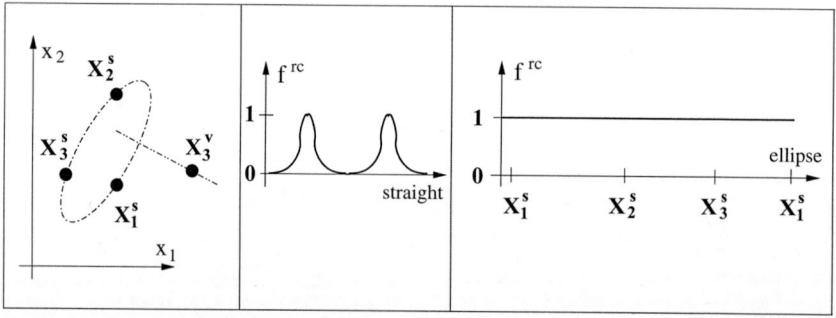

Fig. 3.40. (Left) Ellipse through three seed vectors and perpendicular straight line through a further target vector located far off the ellipse; (Middle) Along the straight line the positive Gaussian course of proximity values is added with the shifted positive Gaussian originating from the target vector, such that the result describes two shifted Gaussians; (Right) Along the ellipse the recognition values are constant 1.

Experiments with the Coarse-to-Fine Strategy of Learning

In the following experiments the coarse-to-fine strategy of learning a recognition function is tested. For the coarse approximation we exemplary use the PCA approach, and the refinement step is based on spherical GBFs. The primary purpose is to obtain a recognition function for a target object of three-dimensional shape, which can be rotated arbitrary and can have different distances from the camera. According to this, both the gray value structure and the size of the target pattern varies significantly. Furthermore, we also deal with a specific problem of appearance-based recognition which arises from inexact pattern localization in the image, *i.e.* shifted target patterns. For assessing the coarse-to-fine strategy of learning we compare the results of recognition with those obtained by a coarse manifold approximation and with those obtained from a 1-nearest-neighbor approach. The role of a first experiment is to obtain an impression of the robustness of recognition if in the application phase the views deviate in several aspects and degrees from the ensembles of seed and validation views. In a second experiment, the effect of the number of seed and validation views is investigated, *i.e.* the role of the dimension of the approximated pattern manifold with regard to robustness of recognition in the application phase.

Experiment Concerning Robustness of Recognition

The first experiment considers three objects, *i.e.* connection box, block of wood, and electrical board. For all three objects the system should learn a recognition function as described above. For this purpose we take 16 images from all three objects, respectively under equidistant turning angles of $22.5°$, altogether 48 images. Figure 3.41 shows a subset of three images from the connection box (in the first row), the block of wood (second row), and the electrical board (third row). For learning the recognition function of the connection box, its 16 images are taken as seed ensemble, and the 32 images from the other two objects are taken as validation ensemble. A similar split-up of the training views takes place for learning the other two recognition functions.

Various Testing Ensembles of Images

For applying the recognition functions we solely consider the connection box and take different testing images from it. For the first set of 16 testing images (denoted by RT_1) the connection box is rotated by an offset angle $8°$ relative to the seed views (see Figure 3.42, first row, and compare with first row in Figure 3.41). For the second set of 16 testing images (denoted by RT_2) the connection box is rotated by an offset angle $14°$ relative to the seed views (see Figure 3.42, second row). For the third set of 16 testing images (denoted by SC_1) the same rotation angles are used as for the seed views, but the distance to the camera has been decreased which results in pattern scaling

Fig. 3.41. Three seed images from a collection of 16 seed images respective for three objects.

factor of about 1.25 (see Figure 3.42, third row, and compare with first row in Figure 3.41). For the fourth set of 16 testing images (denoted by SC_2) the same rotation angles are used as for the seed views, but the distance to the camera has been increased which results in pattern scaling factor of about 0.75 (see Figure 3.42, fourth row). For the fifth set of 16 testing images (denoted by SH_1) the same rotation angles and camera distance are used as for the seed views, but the appearance pattern is shifted by 10 image pixel in vertical direction (see Figure 3.42, fifth row, and compare with first row in Figure 3.41). For the sixth set of 16 testing images (denoted by SH_2) the same rotation angles and camera distance are used as for the seed views, but the appearance pattern is shifted by 10 image pixels in vertical and in horizontal direction (see Figure 3.42, sixth row).

Applying Three Approaches of Recognition

To this testing ensemble of images we apply three approaches of object recognition, denoted by CF_{1NN}, CF_{ELL}, and CF_{EGN}. The approaches have in common that in a first step a testing pattern is projected into three 15-dimensional canonical frames (CFs). These are the eigenspaces of the connection box, the block of wood, and the electrical board, which can be constructed from the 16 seed views, respectively. The second step of the ap-

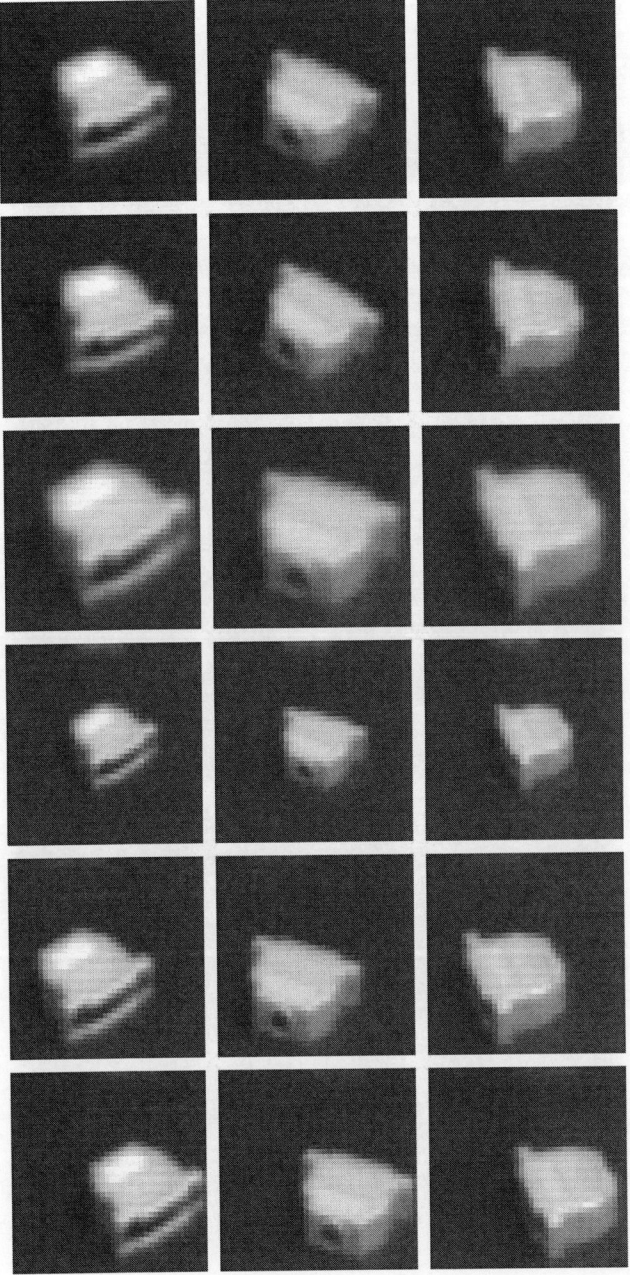

Fig. 3.42. Three testing images from a collection of 16, for six categories respectively.

proaches is the characteristic one. In the first approach CF_{1NN} the recognition of a testing view is based on the maximum proximity to all seed views from all three objects, and the relevant seed view determines the relevant object (it is a *1-nearest-neighbor approach*). In the second approach CF_{ELL} the recognition of a testing view is based on the minimum proximity to the three hyper-ellipsoids constructed from the 16 seed views of the three objects, respectively. Please notice, that all 16 seed views of an object are located on its hyper-ellipsoid, and maybe the testing views deviate only to a certain degree. In the third approach CF_{EGN}, the recognition of a testing view is based on a refinement of the coarse approximation of the pattern manifold by considering counter views with a network of GBFs, *i.e.* the coarse-to-fine approach of learning. The decision for recognition is based on equation (3.37). All three approaches have an equal description length, which is based on the seed vectors of all considered objects, *i.e.* number of seed vectors multiplied by number of components.

Recognition Errors for Various Cases

Table 3.1 contains the numbers of erroneous recognitions for all three approaches and all six sets of testing views. As a first result, we observe that the CF_{ELL} approach yields less errors than the CF_{1NN} approach. Obviously, the three hyper-ellipsoids, which have been constructed from the seed views of the three objects, must describe some appropriate relationships between the seed views. This aspect is completely suppressed in the 1-nearest neighbor approach. As a second result, we observe that the CF_{EGN} approach yields an error-free recognition for the whole testing ensemble and hence surpasses both the CF_{1NN} and the CF_{ELL} approach. This result of the coarse-to-fine strategy of learning is encouraging, because we did take only 16 seed images from the target object and the validation ensemble consisted of counter views from other objects exclusively. The object has been recognized even in case of significant deviations from the seed ensemble of views.

Errors	RT_1	RT_2	SC_1	SC_2	SH_1	SH_2
CF_{1NN}	0	1	4	0	15	16
CF_{ELL}	0	0	0	0	9	1
CF_{EGN}	0	0	0	0	0	0

Table 3.1. Recognition errors for six categories of testing sets by applying three approaches, each testing set consists of 16 elements, 16 seed vectors and 32 validation vectors have been used for learning the recognition functions.

Experiment with Different Sizes of Seed Ensembles

In the second experiment different sizes of seed ensembles are used for learning and applying the three recognition approaches CF_{1NN}, CF_{ELL}, and CF_{EGN}. Once again three objects are considered which look more similar between each other (compared to the objects in the first experiment), *i.e.* integrated circuit, chip carrier, bridge rectifier. Figure 3.43 shows a subset of three images from each object, respectively. In the first case, the seed ensemble of each object may consist of 12 images which are taken under equidistant turning angles of 30°, *i.e.* altogether 36 seed images from all three objects. The split-up between seed and validation ensemble is done for each recognition function according to the approach in the first experiment. We obtain a canonical frame of 11 dimensions, hence the CF_{1NN} approach is working with vectors of 11 components, the CF_{ELL} approach computes distances to hyper-ellipsoids of 11 dimensions, and in the CF_{EGN} the hyper-ellipsoid function is combined with maximal 22 GBF arising from counter views.

Fig. 3.43. Three seed images from three objects, respectively.

Various Testing Ensembles of Images

For applying the recognition functions, we solely consider the integrated circuit and take different testing images from it. For the first set of 12 testing images (denoted by RT) the integrated circuit is rotated by an offset angle 15° relative to the seed views (see Figure 3.44, first row, and compare with first row in Figure 3.43). For the second set of 12 testing images (denoted by SC) the same rotation angles are used as for the seed views, but the distance to the camera has been decreased which results in a pattern scaling factor of about 1.25 (see Figure 3.44, second row, and compare with first row in Figure 3.41). For the third set of 12 testing images (denoted by SH) the same rotation angles and camera distance are used as for the seed views, but the appearance pattern is shifted by 10 image pixel in horizontal direction (see Figure 3.44, third row, and compare with first row in Figure 3.43).

Fig. 3.44. Three testing images for three categories, respectively.

Recognition Errors for Various Cases

Table 3.2 contains the number of erroneous recognitions for all three approaches and all three sets of testing views. The CF_{EGN} approach does not surpass the other two approaches. We explain this unexpected result by the low number of testing views, *i.e.* it is not representative statistically.

Errors	RT	SC	SH
CF_{1NN}	1	2	6
CF_{ELL}	3	0	4
CF_{EGN}	3	0	4

Table 3.2. Recognition errors for three categories of testing sets, each testing set consists of 12 elements, 12 seed vectors and 24 validation vectors have been used for learning the recognition functions.

In the second case, the seed ensemble of each object consists of 20 images which are taken under equidistant turning angles of $18°$, *i.e.* altogether 60 seed images from all three objects. The testing sets RT, SC, and SH are taken according to a strategy similar to the first case but each set consists of 20 images, respectively. Table 3.3 shows that the CF_{EGN} surpasses the CF_{1NN} approach for all three testing sets.

Errors	RT	SC	SH
CF_{1NN}	0	5	11
CF_{ELL}	7	2	7
CF_{EGN}	0	0	8

Table 3.3. Recognition errors for three categories of testing sets, each testing set consists of 20 elements, 20 seed vectors and 40 validation vectors have been used for learning the recognition functions.

In the third case, the seed ensemble of each object consists of 30 images which are taken under equidistant turning angles of $12°$, *i.e.* altogether 90 seed images from all three objects. Once again the CF_{EGN} surpasses the CF_{1NN} approach for all three testing sets as shown in Table 3.4.

Errors	RT	SC	SH
CF_{1NN}	0	8	15
CF_{ELL}	1	1	16
CF_{EGN}	0	0	12

Table 3.4. Recognition errors for three categories of testing sets, each testing set consists of 30 elements, 30 seed vectors and 60 validation vectors have been used for learning the recognition functions.

Finally, in the last case, we use just one testing set and compare the recognition errors directly for different dimensions of the recognition function. In the previous cases of this experiment the relevant testing set has been denoted by SC which consisted of images with rotation angles equal to the seed images, but with a changed distance to the camera. This changed distance to the camera is taken once again, but the whole testing set consists of 180 images from the integrated circuit under rotation with equidistant turning angle of $2°$. Therefore, both a variation of viewing angle and viewing distance is considered relative to the seed views.

Table 3.5 shows the result of applying recognition functions, which have been constructed from 6 seed views (denoted by NS_1), from 12 seed views (denoted by NS_2), from 20 seed views (denoted by NS_3), or from 30 seed views (denoted by NS_4). The result is impressive, because the CF_{ELL} approach clearly surpasses CF_{1NN}, and our favorite approach CF_{EGN} is clearly better than the other two. The course of recognition errors of CF_{EGN}, by increasing the dimension, shows the classical conflict between over-generalization and over-fitting. That is, the number of errors decreases significantly when increasing the dimension from NS_1 to NS_2, but remains constant or even gets worse when increasing the dimension further from NS_2 to NS_3 or to NS_4. Therefore, it is convenient to take the dimension NS_2 for the recognition function as compromise, which is both reliable and efficient. Qualitatively, all our experiments showed similar results.

Errors	NS_1	NS_2	NS_3	NS_4
CF_{1NN}	86	59	50	49
CF_{ELL}	32	3	14	18
CF_{EGN}	24	1	2	3

Table 3.5. Recognition errors for one testing set, which now consists of 180 elements. The approaches of object recognition have been trained alternatively with 6, 12, 20, or 30 seed vectors, for the CF_{EGN} approach we take into account additionally 12, 24, 40, or 60 validation vectors.

As a résumé of all experiments to object recognition, we can draw the conclusion that the dimension of the appearance manifold can be kept surprisingly low.

3.5 Summary and Discussion of the Chapter

The coarse-to-fine approach of learning can be extended and improved in several aspects.

First, in the phase of visual demonstration we can spend more effort in selecting appropriate seed views. For example, trying to determine a probability distribution of views and selecting a subset of the most probable one. In line with this, we can also apply a selection strategy which is based on maximizing the entropy. Furthermore, a support vector approach can be used for selecting from the seed and validation views the most critical one. The common purpose of these approaches is to keep the seed and validation ensemble as small and hence the number of dimensions as low as possible. These approaches belong to the paradigm of *active learning* in which random or systematic sampling of the input domain is replaced by a *selective sampling* [38].

Second, for complicated objects or situations maybe a large set of seed views and/or validation views is necessary for a robust recognition. This set can be splitted up in several subsets and used for constructing canonical frames individually. herefore, the recognition function of an object can be based on a mixture of low-dimensional canonical frames approximating a complicated pattern manifold [165]. Only minor changes need to be done in our approach presented above.

Third, the aspect of temporal continuity in learning and applying recognition functions has been ignored in Subsection 3.4.4, however its important role has been worked out clearly in Subsection 3.4.3. In line with this, it may be useful to consider hyper-ellipsoidal basis functions both for the coarse approximation from the seed views and the fine approximation from validation views.

Fourth, it has already been worked out in Subsections 3.4.1 and 3.4.2 that certain image transformations, such as log-polar transformation, can reduce the complexity of pattern manifolds, and hence reduce the complexity of the recognition function. However, for a reasonable application we must keep a certain relationship between scene object and camera. In the next Chapter 4 this aspect is treated in the context of robotic object grasping using an eye-on-hand system.

Quite recently, novel contributions appeared in the literature which are related to or can further our concept of object recognition. The exploitation of space-time correlations in Subsection 3.4.3 is an exemplary application of the concept of *tangent distance*, as introduced by Simard *et al.* [157]. The work of Arnold *et al.* [8] combines Binford's concept of *quasi-invariants*, the *Lie group analysis*, and *principal component analysis* for constructing *subspace invariants* for object recognition. In Subsection 3.4.4, we presented an approach for learning a recognition function f^{rc} which should be a quasi-invariant for the potential of apprearances of the target object and should discriminate appearances of other objects. Finally, Hall *et al.* [71] presented an approach for merging or splitting Eigenspaces which may contribute to the issue of appropriate dimensionality, as was raised in Subsection 3.4.4.

4. Learning-Based Achievement of RV Competences

For designing and developing autonomous camera-equipped robots this chapter presents a generic approach which is based on systematic experiments and learning mechanisms. The final architecture consists of instructional, behavioral, and monitoring modules, which work on and/or modify vector fields in state spaces.

4.1 Introduction to the Chapter

The introductory section of this chapter embeds our design methodology into the current discussion of how to build behavior-based systems, then presents a detailed review of relevant literature, and finally gives an outline of the following sections.

4.1.1 General Context of the Chapter

Since the early nineties many people were inspired from Brooks' school of behaviorally organized robot systems and followed the underlying ideas in designing autonomous robot systems. Based on diverse experience with implementations, in the late nineties an ongoing discussion started in which advantages and drawbacks of the behavioral robotics philosophy have been weighted up.[1]

Advantages of Brooks' School of Behavior-Based Robotics

The behavioral robotics philosophy originates from the observation of living biological systems whose intelligence can be regarded as layered organization of competences with increasing complexity [25]. Primitive creatures (such as ants) survive with a few low-level competences in which the *reaction* is a dominating characteristic, and sophisticated creatures (such as humans) additionally possess *high level, task-solving competences* in which the *deliberation* is a dominating characteristic. The research community for autonomous

---

[1] For example, online discussion in newsgroup comp.ai.robotics.research in March 1999.

J. Pauli: Learning-Based Robot Vision, LNCS 2048, pp. 171-253, 2001.
© Springer-Verlag Berlin Heidelberg 2001

robots succeeded to synthesize certain low-level competences artificially. The systems work flexible under realistic assumptions, because their experience has been grounded on real environments. By considering a minimalistic aspect, just the information which is relevant for becoming competent must be gathered from the environment. All information acquired from the environment or obtained from internal processes are spread throughout various modules of different control systems on different levels. Both the layered control and the decentralized representation makes the system fault-tolerant and robust in the sense that some vital competences continue to work whereas others can fail. Simple competences, such as obstacle avoiding in a priori unknown environments, can be learned without any user interaction, *i.e.* relying only on sensor information from which to determine rewards or punishments for the learning procedure. In summary, artificial robot systems can acquire and adapt simple behaviors autonomously.

Drawbacks of Brooks' School of Behavior-Based Robotics

In a strict interpretation of Brooks' school the behavioral robotics philosophy can hardly deal with *high-level, deliberate tasks* [167].[2] Deliberations are represented as maps, plans, or strategies, which are the basis for collecting necessary information, taking efficient decisions or anticipating events. Usually, for solving a sophisticated task the necessary information is distributed throughout the whole scene. The information must be collected by moving sensors or cameras and perceiving the scene at different times and/or perspectives. The other aspect of dealing with deliberate tasks is that we are interested in a goal-directed, purposeful behavior in order to solve an exemplary task within the allocated slice of time. Due to the complexity of the particular target goal it must be decomposed in sub-goals and these must be distributed throughout the various modules which achieve certain sub-tasks and thus *contribute to the target goal*. Brooks denies a divide-and-conquer strategy like that and argues that it is difficult if not impossible to explicitly formulate goals or sub-goals. Instead of that, a purposeful overall behavior is supposed to be reached by combining individual behaviors each one working with a generate-and-test strategy which is based on elementary rewards or punishments. However, in our opinion it is extremely difficult to predict the behavior if starting with trial and error, and usually such a strategy proves inadequate to solve an exemplary task at the requested time. In summary, both the lack of a central representation and the lack of purpose injection makes it hard to design an autonomous robot system which is supposed to solve a high-level, deliberate task. The dogmatic adherence to the *strict* behavioral robotics philosophy seems to be not helpful in designing autonomous robots for high-level tasks.

---

[2] We use the attribute *deliberate* for characterizing sophisticated tasks which rely on deliberations and can not be solved in a reactive mode exclusively [34].

Relevant Contributions for Designing Autonomous Robots

We will comment on four relevant contributions in the literature in order to prepare a proposal of a behavior-based design of autonomous robot systems, which saves the advantages and overcomes the drawbacks of the strict behavioral robotics philosophy.

Arkin proposes a *hybrid deliberative/reactive architecture* in which a deliberative component supervises and modifies (if necessary) the reactive component continually [7]. The deliberative component is based on a central representation of information which is subdivided in a short term and a long term memory. The short term memory is used for collecting sensor information that continually comes into and goes out from the sensory range, and the current and past information will be useful for guiding the robot. The long term memory may contain maps from a stable environment or physical parameters from sensors or robot components. *Symbolic planning methods* are proposed for implementing the deliberative components, and *reinforcement learning* is suggested for acquiring reactive behaviors. However, the applicability in hard problems has not been demonstrated. Presumably, the traditional AI methods must be redesigned for real world problems, in which automatic adaptation and learning plays a more fundamental role.

Colombetti *et al.* propose a methodology for designing behavior-based systems which tries to overcome the ad-hocery sometimes criticized in Brooks' system design [39]. The guiding line consists of the following four steps. First, the requirements of the target behavior should be provided. Second, the target behaviors must be decomposed in a system of structured behaviors together with a learning or controlling procedure for each behavior. Third, the structured behaviors should be trained in a simulated or the real environment. Fourth, a behavior assessment takes place based on the pre-specified requirements of the target behavior. In our opinion this design methodology suffers from a vital aspect, namely it is suggested to decompose the target behavior into a system of structured behaviors in a top-down manner. That is, an idealized description of the environment and a description of the robot shell is the basis for the behavioral organization. Furthermore, the control or learning procedures, to be provided for the individual modules, must also be configured and parameterized from abstract descriptions. In our opinion, this designing approach will not work for hard applications, because abstract, idealized decriptions of the real environment are not adequate for a useful design process.

The knowledge-based, top-down design of traditional Artificial Intelligence systems is not acceptable, because the underlying symbolic reasoning mechanisms are lacking robustness and adaptability in practical applications [162]. Instead of that, *experiments* conducted in the real environment must be the design driver, and the design process can only succeed by assessment feedback and modification of design hypotheses in a cyclic fashion. Our favorite methodology is called *behavioral, bottom-up design* to be carried out

during an experimentation phase, in which the nascent robot learns competences by experience with the real environment. This is the only possibility to
become aware of the problems with perception and to realize that perception
must be treated in tough inter-connection with actions. In Subsection 1.3.1
we distinguished two alternative purposes for actuators in camera-equipped
robot systems, *i.e.* dealing with vision-for-robot tasks and robot-for-vision
tasks. Related to these two cases the following ground truths illustrates the
interconnection between perception and action.

> - Visual classes of appearance patterns are relevant and useful only
>   if an actoric verification is possible, *e.g.* a successful robotic grasp-
>   ing is the criterion for verifying hypothetical classes of grasping
>   situations.
> - For visual surveillance of robotic manipulation tasks the camera
>   must be arranged based on perceptual verification which is leading
>   to appropriate observation, *i.e.* successful and non-successful grasps
>   must be distinguishable.

We realize that perception itself is a matter of behavior-based designing
due to its inherent difficulty. One must consider in the designing process the
aspect of computational complexity of perception. This is quite in consensus
with the following statement of Tsotsos [166].

> *"Any behaviorist approach to vision or robotics must deal with the
> inherent computational complexity of the perception problem, other-
> wise the claim that those approaches scale up to human-like behavior
> is easily refuted."*

The computational complexity of perception determines the efficiency and
effectivity of the nascent robot system, and these are two major criteria for
acceptance beside the aspect of autonomy. Usually, any visual or robotic task
leaves some degrees of freedom for doing the perception, *e.g.* choosing among
different viewing conditions and/or among different recognition funtions. It
is a design issue to discover for the application at hand relevant strategies
of making manifold construction tractable, as discussed in Subsection 3.4.1
principally. In line with that, the system design should reveal behaviors which
control the position and orientation of a camera such that it takes on a
normalized relation to scene objects. Under these assumptions, the pattern
manifold for tasks of object recognition or tracking can be simplified, *e.g.* by
log-polar transformation as introduced in Subsection 3.4.2.

4.1.2 Learning Behavior-Based Systems

In Subsections 1.2.2 and 1.3.2 we characterized *Robot Vision* and *Autonomous
Robot Systems* and interconnected both in a definition of *Autonomous Came-*

ra-equipped Robot Systems. This definition naturally combines characteristics of Robot Vision with the non-strict behavioral robotics philosophy, however it does not include any scheme for designing such systems. Based on the discussion in the previous section we can draw the following conclusion concerning the designing aspect.

> The behavior-based architecture of an autonomous camera-equipped robot system must be designed and developed on the basis of experimentation and learning.

Making Task-Relevant Experience in the Actual Environment

In an experimental designing phase the nascent system should make task-relevant experience in the actual environment (see Figure 4.1). An *assessment* is needed of how certain image operators, control mechanisms, or learning procedures behave. If certain constituents do not behave adequately, then a redesign must take place which will be directed and/or supported by a human designer.[3] Simple competences such as obstacle avoiding can be synthesized with minor human interaction mainly consisting of elementary rewards or punishments. Competences for solving high-level, deliberate tasks such as sorting objects can only be learned with intensified human interaction, *e.g.* demonstrating appearances of target objects.

The configuration of competences acquired in the experimental designing phase will be used in the application phase. During the task-solving process a human superviser can observe system behaviors in the real environment. Each contributing behavior is grounded on image operations, actuator movements, feedback mechanisms, assessment evaluations, and learning facilities. For the exceptional case of undesired system behavior, an automatic or manual interruption must stop the system, and a dedicated redesigning phase will start anew. We distinguished between the *experimental designing phase* and the *application phase,* and discussed about system behaviors in both phases. The critical difference is that during the experimental designing phase a human designer is integrated intensively, and during the application phase just a human supervisor is needed for interrupting the autonomous system behaviors in exceptional cases.

Task-Relevant Behaviors as a Result of Environmental Experience

The outcome of the experimental designing phase is a configuration of task-relevant behaviors (see again Figure 4.1). The task-relevance is grounded on making purposive experiences in the actual environment in order to become acquainted with the aspects of situatedness and corporeality (see Subsection

---

[3] In our opinion, the automatic, self-organized designing of autonomous robot systems for solving high-level, deliberate tasks is beyond feasibility.

1.3.2). We have already worked out in Chapters 2 and 3 that *visual demonstration and learning processes* play the key role in acquiring operators for object localization and recognition. The acquisition and/or fine-tune of image operators is typically done in the experimentation phase. Generally, image operators are a special type of behaviors which extract and represent task-relevant information in an adaptive or non-adaptive manner. Those *image processing behaviors* are constituents of robotic behaviors which are responsible for solving robotic tasks autonomously. Other constituents of robotic behaviors may consist of strategies for scene exploration, control procedures for reaching goal situations, learning procedures for acquiring robot trajectories, *etc.* It is the purpose of the designing phase to decompose the target behavior into a combination of executable behaviors, determine whether a behavior should be *adaptive or non-adaptive*, and specify all relevant constituents of the behaviors. For the behavioral organization we have to consider requirements of robustness, flexibility, and time limitation, simultaneously.

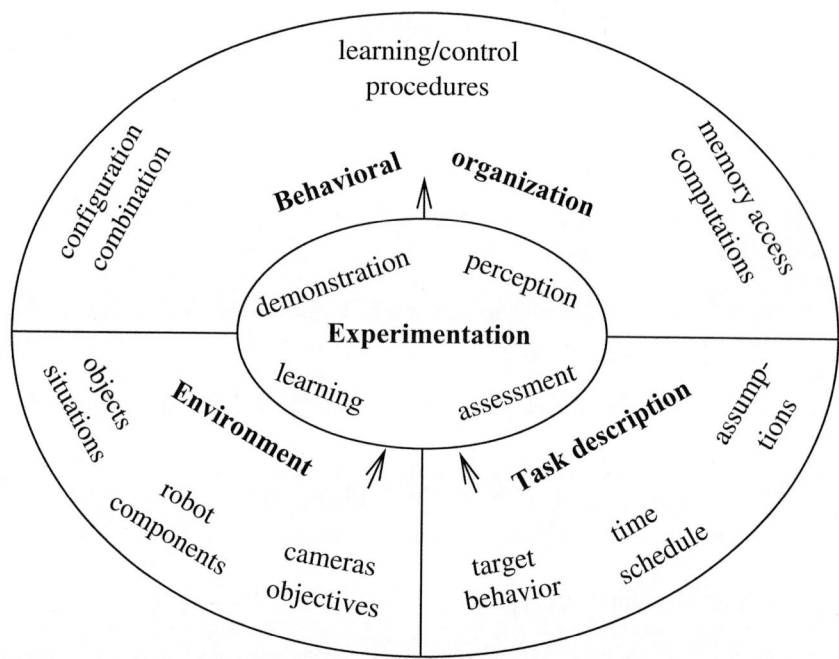

Fig. 4.1. Central role of experimentation for achieving autonomous robots.

Image-Based Robot Servoing

The backbone of an autonomous camera-equipped robot system consists of mechanisms for *image-based robot servoing*. These are continual processes of *perception-action cycles* in which actuators are gradually moved and continually controlled with visual sensory feedback. Primary applications are *tool handling* of manipulating objects and *camera movement* for detailed, incremental object inspection. For example, for the purpose of object inspection the camera head should be controlled to reach a desired size, resolution, and orientation of the depicted object. A secondary application is the *self-characterization of cameras*, *i.e.* determining the optical axes, the fields of view, and the location of the camera head. In the experimental designing phase one has to configure and parameterize servoing mechanisms, *e.g.* redesign non-successful controllers and find out successful controllers which reach their sub-goals.

We are convinced that the potential usefulness of image-based robot servoing is far from being sufficiently realized which is due to several reasons. First, visual goal situations of robot servoing should be grounded in the real environment, *i.e.* should be specified by visual demonstration in the experimental designing phase. For example, in Subsection 3.3.4 we demonstrated examples of scored grasping situations, which can be used as guiding line for servoing the robot hand towards the most stable grasping position. Second, the various *degrees-of-freedom (DOF)* of a robot head, *e.g.*, pan, tilt, vergence, focus, focal length, and aperture of the head-cameras, must be controlled in cooperation in order to exploit their complementary strengths. Unfortunately, our work will not contribute to this problem. Third, nearly all contributions to robotic visual servoing describe systems consisting of just one robot, *e.g.*, exclusively a robot manipulator or a robot head. Instead of that, we present applications of image-based robot servoing for a *multi-component robot system* consisting exemplary of a *mobil robot head*, a *stationary manipulator*, and a *rotary table*.

The Issue of a Memory in Behavior-Based Systems

A memory will be put at the disposal of the behavior-based camera-equipped robot system. This is inevitable necessary, if high-level, deliberate tasks must be solved autonomously and requirements like robustness, flexibility, and time limitation should hold simultaneously. It is convenient to decentralize the memory according to certain competences of the task-solving process, *i.e.* introducing a *shared competence-centered memory*. The memory of the system preserves the opportunity for incremental development, because the need for frequent sensing in stable environments is reduced and information that is outside of the sensory range can be made use of for guiding the robot. The discovery of movement strategies for actuators will be accelerated based on memorized bias information which can be acquired in previous steps of the task-solving process. For a high-level task-solving strategy several behaviors contribute information at different phases. It is this characteristic of *contributory behaviors* along with the diverse flow of information which lets us prefer a shared memory technology (as opposed to a message passing technology) and thus simplify the designing of the behavior-based robot system.

Acquiring Knowledge by Interaction with the Environment

As opposed to traditional knowledge-based systems, all memory contents of a behavior-based system must be acquired and verified by *system interaction with the environment*. In the experimental designing phase this interaction is directed by a human system designer, and in the application phase the interaction should be autonomous. During the experimental designing phase the *basic memory stuff* will be acquired, *e.g.* feature compatibilities for object localization, pattern manifolds for object recognition, parameters for image pre-processing and feature extraction, camera-related characteristics like geometric relation to the robot, optical axis, and field of view, parameters of visual feedback mechanisms, maps or strategies of camera movement for scene inspection, or of arm movement for object handling, *etc.* In the application phase the *hot memory stuff* for solving the task will be acquired, *e.g.* positions of obstacles to be localized by moving the camera according to a fixed or exploratory strategy, obstacle-avoiding policy of manipulator movement for approaching a target position, geometric shape of a target object to be determined incrementally by observation from different viewpoints, *etc.* During this application phase various system constituents will read from and write to certain regions of the memory, *i.e.* have a share in the memory. We summarize this discussion as follows.

A camera-equipped robot system can reach autonomy in solving certain high-level, deliberate tasks if three preconditions hold.

- A behavioral, experimental designing phase must take place in the actual environment under the supervision of a system designer.
- In the application phase various mechanisms of image-based robot servoing must be available for reaching and/or keeping sub-goals of the task-solving process.
- Various exploration mechanisms should incrementally contribute information for the task-solving process and continually make use of the memorized information.

The following section reviews relevant contributions in the literature.

4.1.3 Detailed Review of Relevant Literature

The behavioral, experimental designing phase is composed of demonstration, perception, learning, and assessment. The important role of this designing methodology for obtaining operators for object localization and recognition has already been treated in Section 1.4 and in Chapters 2 and 3, extensively. In this chapter the focus is more on acquiring strategies for decomposing a target behavior into a configuration of simpler behaviors, on the interplay between camera movement and object recognition or scene exploration, and on the learning of purposive perception-action cycles. The tight coupling between perception and action has also been emphasized by Sommer [162]. A more detailed treatment is presented in a work of Erdmann [51] who suggests to design so-called *action-based sensors*, *i.e.* sensors should not recognize states but *recognize applicable actions*. A tutorial overview to action selection mechanisms is presented by Pirjanian [129, pp. 17-56], including priority-base arbitration (*e.g.* Brooks' subsumption architecture), state-based arbitration (*e.g.* reinforcement learning approach), winner-take-all arbitration (*e.g.* activation networks), voting-based command fusion (*e.g.* action voting), fuzzy-command fusion, superposition-based command fusion (*e.g.* dynamical systems approach).

A special journal issue on *Learning Autonomous Robots* has been published by Dorigo [49, pp. 361-505]. Typical learning paradigms are supervised learning, unsupervised learning, reinforcement learning, and evolutionary learning. With regard to applying video cameras only two articles are included which deal with automatic navigation of mobile robots. We recommend these articles to the reader as bottom-up designing principles are applied, *e.g.* learning to select useful landmarks [67], and learning to keep or change steering directions [13]. Another special journal issue on *Robot Learning* has been published by Sharkey [155, pp. 179-406]. Numerous articles deal with reinforcement learning which are applied in simulated or real robots. Related to the aspect of bottom-up designing the work of Murphy *et al.* [113] is interesting, because it deals with learning by experimentation especially for determining navigational landmarks. A special journal issue on applications of reinforcement learning is also published by Kaelbling [88] and a tutorial introduction is presented by Barto *et al.* [14].

The paradigm *Programming by Demonstration (PbD)* is relevant for this chapter. Generally, a system must record user actions and based on that must generalize a program that can be used for new examples/problems. Cypher edited a book on this topic which contains a lot of articles reporting about methodologies, systems and applications [45]. However, no articles on automated generation of vision algorithms or robot programs are included. Instead of that, the work of Friedrich *et al.* applies the PbD paradigm to generating robot programs by generalizing from several example sequences of robot actions [58]. The approaches in [84] and [93] differ from the previous one especially in the aspect that a human operator demonstrates objects, situations, actuator movements, and the system should make observations in order to recognize the operator intention and imitate the task solution. All three systems focus on the symbolic level and treat image processing or visual feedback mechanisms only to a little extent.

We continue with reviewing literature on image-based robot servoing. Quite recently, a special issue of the International Journal on Computer Vision has been devoted to Image-based Robot Servoing [79]. A book edited by Hashimoto gives an overview of various approaches of automatic control of mechanical systems using visual sensory feedback [76]. Let us just mention the introductory, tutorial work of Corke [41].[4] There, two approaches of *visual servoing* are proposed, *i.e.* the *position-based* and the *feature-based* approach. In position-based control features are extracted from the image and used in conjunction with a geometric model of the target to determine the pose of the target with respect to the camera. In image-based servoing the last step is omitted, and servoing is done on the basis of image features directly. In our applications we use both approaches depending on specific sub-tasks.

Hager *et al.* describe a system that positions a robot manipulator using visual information from two stationary cameras [70]. The end-effector and the visual features defining the goal position are simultaneously tracked using a *proportional–integral (PI) controller*. We adopt the idea of using *Jacobians* for describing the 3D–2D relation but taking projection matrices of a poorly calibrated head-camera–manipulatior relation into account instead of explicit camera parameters.

The system of Feddema *et al.* tracks a moving object with a single camera fastened on a manipulator [54]. A visual feedback controller is used which is based on an inverse Jacobian matrix for transforming changes from image coordinates to robot joint angles. The work is interesting to us because the role of a *teach-by-showing method* is mentioned. Offline the user teaches the robot desired motion commands and generates *reference vision-feature data*. In the online playback mode the system executes the motion commands and controls the robot until the extracted feature data correspond to the reference data.

Papanikolopoulos and Khosla present an algorithm for robotic camera servoing around a static target object with the purpose of reaching a certain relation to the object [124]. This is done by moving the camera (mounted on a manipulator) such that the perspective projections of certain feature points of the object reach some desired image positions. In our work, a similar problem occurs in controlling a manipulator to carry an object towards the head-camera such that a desired size, resolution, and orientation of the depicted object is reached.

We continue with reviewing literature on mechanisms for *active scene exploration* and *object inspection* including techniques for incremental information collection and representation. A book edited by Landy *et al.* gives the state of the art of *exploratory vision* and includes a chapter on robots that explore [98]. Image-based robot servoing must play a significant role especially in model-free exploration of scenes. A typical exploration technique for completely unknown environments is reinforcement learning which has already been mentioned above.

The two articles of Marchand and Chaumette [104] and Chaumette *et al.* [35] deal with 3D structure estimation of geometric primitives like blocks

---

[4] A tutorial introduction to visual servo control of robotic manipulators has also been published by Hutchinson *et al.* [82].

and cylinders using active vision. The method is based on controlled camera motion of a single camera and involves to gaze on the considered objects. The intention is to obtain a high accuracy by focusing at the object and generating optimal camera motions. An optimal camera movement for reconstructing the cyclinder would be a cycle around it. This camera trajectory is acquired via visual servoing around a cylinder by keeping the object depiction in vertical orientation in the image center. A method is presented for combining 3D estimations under several viewpoints in order to recover the complete spatial structure. Generally, the gaze planning strategy mainly uses a representation of known and unknown spatial areas as a basis for selecting viewpoints.

A work of Dickinson *et al.* presents an active object recognition strategy, which combines the use of an attention mechanism for focusing the search for a 3D object in a 2D image with a viewpoint control strategy for disambiguating recovered object features [47]. For example, the sensor will be servoed to a viewing position such that different shapes can be distinguished, *e.g.* blocks and cylinders. Exploration techniques play also a role in tasks of vision-based, robotic manipulations like robotic grasping. Relevant grasp approaching directions must be determined which can be supported by training the system to grasp objects. Techniques of active learning will reduce the number of examples from which to learn [144]. The emphasis of this chapter is not on presenting sophisticated approaches of scene exploration, but on the question of how to integrate such techniques in a behavioral architecture.

Exploration mechanisms must deal with target objects, obstacle objects, and navigational strategies. A unified framework for representation can be provided by so-called *dynamic vector fields* which are defined in the *dynamic systems theory* [153, 50]. *Attractor vector fields* are virtually put at the positions of target objects, and *repellor vector fields* are virtually placed at the positions of obstacle objects. By summarizing all contributing vector fields we obtain useful hypotheses of goal-directed, obstacle-avoiding navigation trajectories. In a work of Mussa-Ivaldi the attractors and repellors are defined exemplary by radial basis functions and gradients thereof [115]. This chapter will make use of the mentioned methodology for visually navigating a robot arm through a set of obstacle objects with the task of reaching and grasping a target object.

Generally, the mentioned kind of vector fields simulate attracting and repelling forces against the robot effectors and therefore can be used for planning robot motions. The vector fields are obtained as gradients of so-called *potential functions* which are centered at target and obstacle positions. A tutorial introduction to so-called *potential field methods* is presented by Latombe [99, pp. 295-355]. In this chapter, we use the term *dynamic vector field* in order to emphasize the *dynamics* involved in the process of solving a high-level task. The sources of dynamics are manifold. For example, sub-goals are pursued successively, and certain situations must be treated unexpectedly. Furthermore, coarse movement plans are refined continually based on latest scene information acquired from images. Related to that, *gross motions* are performed to roughly approach targets on the basis of coarse reconstructions and afterwards *fine motions* are performed to finally reach haptic contact [99, pp. 452-453]. The system dynamics will be reflected in ongoing changes of involved vector fields.

The dynamic vector field approach considers the inherent inaccuracy of scene reconstruction naturally. This is achieved by specifying the width (support) of the underlying potential functions dependent on expected levels of inaccuracy. The related lengths of the gradient vectors are directly correlated with acceptable minimal distances from obstacles. Another advantage is the *minimal disturbance principle, i.e.* a local addition or subtraction of attractors or repellors causes just a local change of the movement strategy. However, this characteristic of locality is also the reason for difficulties in planning global movement strategies. Therefore, apart from potential field methods, many other approaches on robot motion planning have been developed [99]. For example, graph-based approaches aim at representing the global connectivity of the robot's free space as a graph that is subsequently searched for a minimum-cost path. Those approaches are favourable for more complicated planning problems, e.g. generating assembly sequences [141]. This chapter is restricted to simple planning problems for which dynamic field methods are sufficient.

The *deliberative* task of robot motion planning must be interleaved with the *reactive* task of continual feedback-based control. In the work of Murphy *et al.* [114] a planning system precomputes an a priori set of optimal paths, and in the online phase terrain changes are detected which serve to switch the robot from the current precomputed path to another precomputed path. A real-time fusion of deliberative planning and reactive control is proposed by Kurihara *et al.* [95], whose system is based on a cooperation of behavior agents, planning agents, and behavior-selection agents. The system proposed by Donnart and Meyer [48] applies reinforcement learning to automatically acquire planning and reactive rules which are used in navigation tasks. Unfortunately, most of the mentioned systems seem to work only in simulated scenes. Beyond the application in real scenes, the novelty of this chapter is to treat the deliberative and the reactive task uniformly based on the methodology of dynamic vector fields.

4.1.4 Outline of the Sections in the Chapter

In Section 4.2 several basis mechanisms are presented, such as visual feedback control, virtual force superposition, and integration of deliberate strategies with visual feedback. A series of generic modules for designing instructions, behaviors, and monitors is specified. Section 4.3 introduces an exemplary high-level task, discusses designing-related aspects according to the bottom-up methodology, and presents task-specific modules which are based on the generic modules of the preceding section. In Section 4.4 modules for acquiring the coordination between robots and cameras are presented. Depending on the relevant controller, which is intended for solving a sub-task, we present appropriate mechanisms for estimating the head-camera–manipulator relation. Furthermore, image-based effector servoing will be applied for determining the optical axis and the field of view of the head-camera. Section 4.5 discusses the approaches of the preceding sections.

4.2 Integrating Deliberate Strategies and Visual Feedback

We describe two basis mechanisms of autonomous camera-equipped robot systems, i.e. virtual force superposition and visual feedback control.[5] Both mechanisms can be treated cooperatively in the framework of so-called *deliberate and reactive vector fields*. Different versions and/or combinations of the basis mechanisms occur in so-called *basic, generic modules* which can be used as library for implementing *task-specific modules*. As a result of several case studies on solving high-level, deliberate tasks (using autonomous camera-equipped robot systems) we have discovered 12 different categories of generic modules. They are subdivided in three categories of *instructional modules*, six categories of *behavioral modules*, and three categories of *monitoring modules*. Task-specific modules make use of the basic, generic modules, but with specific implementations and parametrizations. In the second part of this section we present and explain the scheme of the task-specific modules and the basic, generic modules. The basis mechanisms and the generic modules will be applied to an exemplary high-level task in Section 4.3.

4.2.1 Dynamical Systems and Control Mechanisms

State space of proprioceptive and exteroceptive features In Subsection 1.3.2 we introduced the criteria situatedness and corporeality for characterizing an autonomous robot system and especially we treated the distinction between *proprioceptive* and *exteroceptive* features. Proprioceptive features describe the state of the effector, and exteroceptive features describe aspects of the environmental world in relation to the effector.

Transition Function for States

The proprioceptive feature vector of the effector is subdivided in a *fixed state vector S^c* and a *variable state vector $S^v(t)$*. The vector S^c is inherent constant, and the vector $S^v(t)$ can be changed through a *vector of control signals $C(t)$*

---

[5] For solving high-level, deliberate tasks the camera-equipped robot system must be endowed with further basic mechanisms, *e.g.* reinforcement learning and/or unsupervised situation clustering. However, a detailed treatment is beyond the scope of this work.

at time t. For example, the fixed state vector of a robot manipulator contains the *Denavit-Hartenberg parameters length, twist, offset* for each link which are constant for rotating joints [42]. The variable state vector $S^v(t)$ could be the 6-dimensional state of the robot hand describing its position (in 3D coordinates X, Y, Z) and orientation (in Euler angles *yaw, pitch, roll*). On the basis of the variable state vector $S^v(t)$ and control vector $C(t)$ the *transition function* f^{ts} determines the next state vector $S^v(t + 1)$.

$$S^v(t + 1) := f^{ts}(C(t), S^v(t)) \tag{4.1}$$

If the vectors $C(t)$ and $S^v(t)$ are of equal dimension with the components corresponding pairwise, and the function f^{ts} is the vector addition, then $C(t)$ serves as an increment vector for $S^v(t)$. For example, if the control vector for the robot hand is defined by $C(t) := (\triangle X, \triangle Y, \triangle Z, 0, 0, 0)^T$, then after the movement the state vector $S^v(t+1)$ describes a new position of the robot hand preserving the orientation. Both the state and control vector are specified in the manipulator coordinate system.

Task-relevant control vectors are determined from a combination of proprioceptive and exteroceptive features. From a system theoretical point of view a camera-equipped robot system is a so-called *dynamical system* in which the geometric relation between effectors, cameras, and objects changes dynamically. The process of solving a high-level, deliberate task is the *target behavior of the dynamical system*, which can be regarded as a journey through the state space of proprioceptive and exteroceptive features. The characteristic of the task and the existence of the environment (including the robot) are the basis for relevant affinities and constraints from which to determine possible courses of state transitions.

Virtual Force Superposition

Affinities and constraints can be represented as *virtual forces* uniquely. The trajectory of a robot effector is determined by a *superposition* of several virtual forces and discovering the route which consumes minimal energy and leads to a goal state. This goal state should have the characteristic of *equilibrium* between all participating forces. For example, the task of robotic navigation can be regarded as a mechanical process in which the target object is virtually attracting the effector and each obstacle object is virtually repelling the effector. *Attractor forces* and *repellor forces* are the basic constituents determining the overall behavior of an effector in a dynamical system.

Attractor and Repellor Vector Fields

We define *attractor vector fields* and *repellor vector fields* for the space of variable state vectors. Let S_A^v denote a particular state vector, which is taken as the position of an attractor. Then, the attractor vector field is defined by

$$VF_A[S_A^v](S^v) := v_A \cdot (S_A^v - S^v) / \parallel S_A^v - S^v \parallel \tag{4.2}$$

The force vectors at all possible state vectors S^v of the effector are oriented to the attractor position S_A^v and factor v_A is specifying the unique length of the force vectors (see left image in Figure 4.2). On the other hand, a particular state vector S_R^v is taken as the position of a repellor. The repellor vector field can be defined by

$$VF_R[S_R^v](S^v) := 2 \cdot v_B \cdot (S^v - S_R^v) \cdot \exp\left(- \parallel \beta \cdot (S^v - S_R^v) \parallel^2\right) \quad (4.3)$$

The force vectors in the neighborhood of the repellor position are directed radially off this position, and the size of the neighborhood is defined by factor v_B (see right image in Figure 4.2).

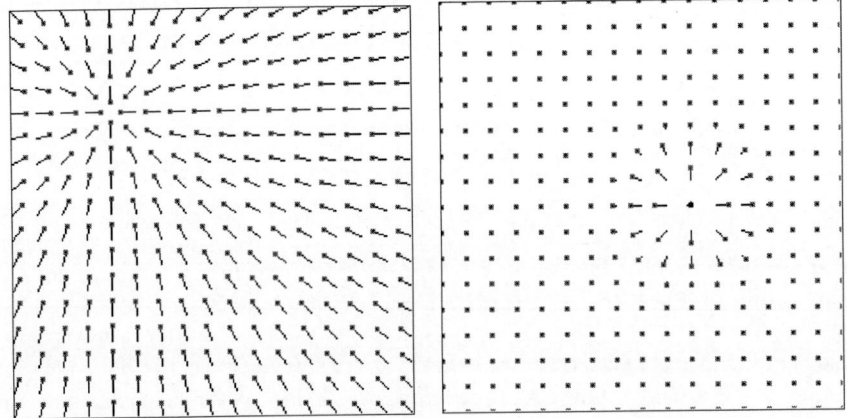

Fig. 4.2. (Left) Attractor vector field; (Right) Repellor vector field.

Trajectories in the Superposed Vector Fields

The overall force vector field is simply obtained by summing up the collection of attractor vector fields together with the collection of repellor vector fields.

$$VF_O[\{S_{Ai}^v\}, \{S_{Rj}^v\}](S^v) := \sum_{S_{Ai}^v} VF_A[S_{Ai}^v](S^v) +$$

$$\sum_{S_{Rj}^v} VF_R[S_{Rj}^v](S^v) \quad (4.4)$$

For example, the left image in Figure 4.3 shows the superposition of the attractor and the repellor vector field of Figure 4.2. With these definitions the control vector $C(t)$ can be defined subject to the current state vector $S^v(t)$ of the effector by

$$C(t) := VF_O[\{S_{Ai}^v\}, \{S_{Rj}^v\}](S^v(t)) \quad (4.5)$$

A trajectory towards the state vector S_A^v under by-passing state vector S_R^v is shown in the right image of Figure 4.3. It has been determined by applying equation (4.1) iteratively.

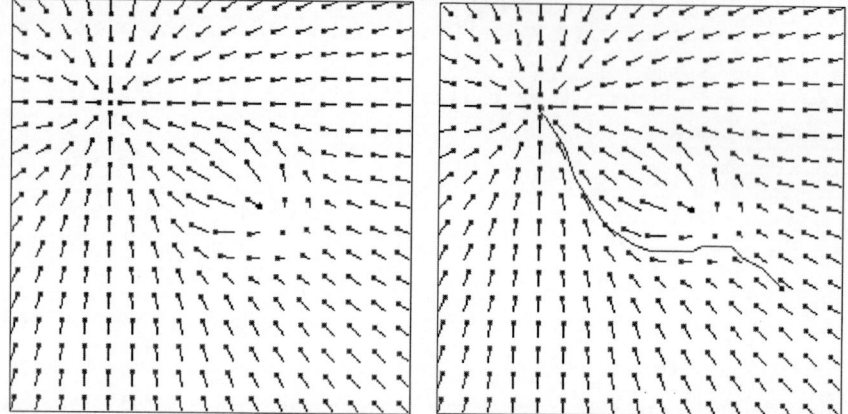

Fig. 4.3. (Left) Superposition of attractor and repellor vector field; (Right) Trajectory towards state vector S_A^v under by-passing state vector S_R^v.

It is convenient to call an attractor vector field simply an *attractor* and to call a repellor vector field simply a *repellor*. Furthermore, we call an equilibrium point of a vector field simply an *equilibrium*. The force vector length at an equilibrium is zero, *i.e.* no force is at work at these points. For example, an equilibrium is located at the center of an attractor vector field (see left image of Figure 4.2). On the other hand, a repellor vector field also contains an equilibrium at the center but additionally an infinite set of equilibriums outside a certain neighborhood (see dots without arrows in the right image of Figure 4.2).

Vector Fields for Representing Planned and Actual Processes

The force vector field can serve as a unique scheme both for representing a high-level task and for representing the task-solving process which the camera-equipped robot system is involved in. As a precondition for enabling situatedness and corporeality, the system must determine attractors and repellors automatically from the environmental images. The critical question is: *Which supplements of the basic methodology of force vector fields are necessary in order to facilitate the treatment of high-level, deliberate tasks ?* In the following subsection we introduce visual feedback control and explain why this supplement is necessary. After this, a distinction between deliberate and reactive vector fields is introduced for organizing the deliberate and reactive aspects of solving a high-level task.

Generic Mechanism of Visual Feedback Control

The particular state vectors S_A^v in equation (4.2) and S_R^v in equation (4.3), based on which to specify attractors and repellors, are represented in the same vector space as the *variable state vector* $S^v(t)$ of the effector. For example, for the task of robotic object grasping the vector $S^v(t)$ represents the robot hand position, vector S^v_A represents the target object position, and S^v_R represents an obstacle object position, and all these positions are specified in a common coordinate system which is attached at the manipulator basis. Through control vector $C(t)$ the state vector $S^v(t)$ of the robot effector is changing and so does the geometric relation between robot effector and target object.

For interacting with the environment the vectors S^v_A and S^v_R must be determined on the basis of exteroceptive features, *i.e.* they are extracted from images taken by cameras. The acquired vectors are inaccurate to an unknown extent because of inaccuracies involved in image processing. With regard to the variable state vector $S^v(t)$ we must distinguish two modalities. On the one hand, this current state of the robot effector is computed simply by taking the forward kinematic of the actuator system into account. On the other hand, the state of the robot effector can be extracted from images if the effector is located in the field of view of some cameras. Of course, for comparing the two results we must represent both in a unique coordinate system. However, the two representations are not equal, which is mainly due to friction losses and inaccuracies involved in reconstruction from images.

In summary, all vectors S^v_A, S^v_R, and $S^v(t)$, acquired from the environment, are inaccurate to an unknown extent. Consequently, we will regard the force vector field just as a bias, *i.e.* it will play the role of a backbone which represents planned actions. In many sub-tasks it is inevitable to combine the force vector field with a control mechanism for fine-tuning the actions based on *continual visual feedback*. Based on this discussion we introduce the methodology of image-based effector servoing.

Definition 4.1 (Image-based effector servoing) *Image-based effector servoing is the gradual effector movement of a robot system continually controlled with visual sensory feedback.*

Measurement Function and Control Function

In each state of the effector the cameras take images from the scene. This is symbolized by a *measurement function* f^{ms} which produces a *current measurement vector* $Q(t)$ at time t (in coordinate systems of the cameras).

$$Q(t) := f^{ms}(S^v(t), S^c) \tag{4.6}$$

The current state vector $S^v(t)$ in equation (4.6) is supposed to be determined by *forward kinematics*. The current measurement vector $Q(t)$ may also contain features which are based on the fixed state vector of the effector but are extracted from image contents, *e.g.* image features which describe the appearance of the gripper fingers.

According to this, the two modalities of representing the current state of the effector (as discussed above) are treated in the equation. Given the current measurement vector $Q(t)$, the current state vector $S^v(t)$, and a *desired measurement vector* Q^*, the controller generates a control vector $C(t)$.

$$C(t) := f^{ct}(Q^*, Q(t), S^v(t)) \qquad (4.7)$$

The *control function* f^{ct} describes the relation between changes in different coordinate systems, *e.g.*, $Q(t)$ in the image and $S^v(t)$ in the manipulator coordinate system. The control vector $C(t)$ is used to update the state vector into $S^v(t+1)$, and then a new measurement vector $Q(t+1)$ is acquired which is supposed to be more closer to Q^* than $Q(t)$. In the case that the desired situation is already reached after the first actuator movement, the *one-step controller* can be thought of as an exact *inverse model* of the robot system. Unfortunately, in realistic control environments only approximations for the inverse model are available. In consequence of this, it is necessary to run through cycles of gradual actuator movement and continual visual feedback in order to reach the desired situation step by step, *i.e. multi-step controller*.

Offline- and Online-Phase of Image-Based Effector Servoing

Image-based effector servoing is organized into an offline-phase and an online-phase. Offline we specify the approximate *camera–manipulator relation* of coordinate systems and define the control function f^{ct} thereof. Frequently, the control function is a *linear approximation* of the unknown inverse model, *i.e.*, the parameters $Q^*, Q(t), S^v(t)$ are linear combined to produce $C(t)$.[6] Online the control function is applied during which the system recognizes a current situation and compares it with a certain goal situation. In case of deviation the effector is moving to bring the new situation closer to the goal situation. This cycle is repeated until a certain threshold criterion is reached.

It is characteristic for image-based effector servoing to work with current and desired measurements $Q(t)$ and Q^* in the image directly and avoid an explicit reconstruction into the coordinate system of the effector. Typically, these image measurements consist of 2D position and appearance features which suffice in numerous applications and need not to be reconstructed. For example, without reconstructing the 3D object shape it is possible to control the viewing direction of a camera such that the optical axis is directed to the center of the object silhouette.

Image-Based Servoing in the Framework of Force Vector Fields

Image-based effector servoing can be regarded conveniently in the framework of force vector fields. However, the basic vector space does not consist of effector state vectors $S^v(t)$, but consists of image measurement vectors $Q(t)$. Effector servoing is goal-directed, as defined by equations (4.1), (4.6), (4.7), and can be represented by a fairly simply force vector field. It is just an

---

[6] Some articles in a book edited by Hashimoto [76] also describe nonlinear, fuzzy logic, and neural network control schemes.

attractor vector field, with one attractor specified by the desired measurement vector Q^*, and no repellors are involved. The trajectory from $Q(t)$ towards Q^* is supposed to be a straight course, however it will be a jagged course actually. This is because the current measurement vector $Q(t)$ can be changed only indirectly by effector movement, and the relevant control function f^{ct} is just an approximation which is inaccurate more or less.

Compromise between Plan Fulfillment and Plan Adjustment

In summary, in high-level, deliberate tasks we have to deal with two types of force vector fields, the first one has been defined in terms of effector state vectors and the second one in terms of image measurement vectors. This can also be realized in the two equations (4.5) and (4.7) which specify two different definitions for control vector $C(t)$. The first one is obtained from a force vector field which will play the role of a plan for an effector trajectory. The second one is used for locally changing or refining the effector trajectory and thus is responsible for deviating from or adjusting the plan, if necessary according to specific criteria. In other words, the first category of force vector fields (which we call *deliberate vector fields*) is responsible for the deliberate aspect and the second category of force vector fields (which we call *reactive vector fields*) for the reactive aspect of solving high-level robotic tasks.

In general case, one must find compromise solutions of plan fulfillment and plan adjustment. However, depending on the characteristic of the task and of the environment there are also special cases, in which either the reactive or the deliberate aspect is relevant exclusively. That is, image-based effector servoing must be applied without a deliberate plan, or on the other hand, a deliberate plan will be executed without frequent visual feedback. We present examples for the three cases later on in Subsection 4.2.2 and in Section 4.3.

Managing Deliberate and Reactive Vector Fields

The overall task-solving process of a camera-equipped robot system can be decomposed in so-called elementary and assembled behaviors. It will prove convenient to introduce an *elementary behavior* as the process of image-based effector servoing by which the current measurement vector is transformed iteratively into a desired measurement vector. Specifically, the basic representation scheme of an elementary behavior is a reactive vector field constructed by just one attractor, and the accompanying equilibrium must be approached by continual visual feedback control of the effector. The control vector $C(t)$ is defined according to equation (4.7). In addition to this, we introduce an *elementary instruction* as an atomic step of changing effector state $S^v(t)$ without visual feedback. The control vector $C(t)$ can be defined according to equation (4.5). An *assembled instruction* is a sequence of elementary instructions to be strung together, *i.e.* it is a course of effector movements as shown in the right image of Figure 4.3 exemplary. Elementary and assembled instructions are determined according to a plan which is represented as a deliberate vector field. Finally, we define an *assembled behavior* as the composition of at least one elementary behavior with further elementary behaviors and/or elementary or assembled instructions. Several combinations are conceivable depending on requirements of the application. A behavior (elementary or assembled) does include *perception-action cycles*, and an instruction (elementary or assembled) does not.

Instructions, Behaviors, Sub-tasks, Vector Fields

The instructions and behaviors of a task-solving process reflect the decomposition of a high-level task into several sub-tasks. It will prove convenient to distinguish between elementary sub-tasks and assembled sub-tasks, with the latter being composed of the former. For solving an *elementary sub-task* one is working with a partial short-term plan, or is working without a plan at all. In the first case, in which a plan is involved, we will stipulate that in an elementary sub-task the camera-equipped robot system is trying to reach just one deliberate goal and perhaps must keep certain constraints (constrained goal achievement). Related to the methodology of dynamical systems the overall force vector field for a sub-task must be constructed by just one attractor vector field and optionally summing it up with a collection of repellor vector fields, *i.e. one attractor and optionally several repellors*. The natural restriction on one attractor in a deliberate vector field reduces the occurrence of ambiguities while planning effector courses (*e.g.* only one equilibrium should occur in the whole vector field).[7] The second case in which no plan is involved means that an elementary sub-task must be solved by an elementary behavior (as introduced above) or any kind of combination between several elementary behaviors. However, no deliberate support is provided, *i.e.* trivially the deliberate vector field is empty without any virtual force vectors. The actual effector movements are controlled by visual feedback exclusively which is represented in reactive vector fields.

Generic Deliberate Vector Fields, Current Deliberate Vector Fields

During the task-solving process certain elementary sub-tasks will come to completion continually and other ones must be treated from the beginning, *i.e. assembled sub-tasks*. Related to the planning aspect a short-term plan is replaced by another one and all of which belong to an overall long-term plan. Related to the methodology of dynamical systems the *deliberate vector field is non-stationary* during the overall process of task-solution. This means, from elementary sub-task to elementary sub-task both the attractor and the collection of repellors change in the overall vector field according to the replacement of the short-term plan. We introduce the following strategy for handling this *non-stationarity*. A so-called *generic deliberate vector field* is

---

[7] Opposed to that, the overall force vector field defined in equation (4.4) consists of several attractors and repellors including several equilibrium points.

constructed which represents the overall task, and from that we dynamically construct so-called *current deliberate vector fields* which are relevant just at certain intervals of time. Each interval is the time needed to solve a certain elementary sub-task.

Examples for the Two Categories of Deliberate Vector Fields

A scene may consist of a set of target objects which should be collected with a robot arm. A useful strategy is to collect the objects in a succession such that obstacle avoiding is not necessary. The generic vector field is generated by constructing attractors from all target objects. The current vector field should consist of just one attractor and will be derived from the generic field continually. The relevant succession of current vector fields is obtained based on considering the geometric proximity and relation between the target objects and the robot arm. Figure 4.4 shows in the left image the process of approaching the effector to the accompanying equilibrium of the first attractor (constructed at the first target object). Then, the object will be removed which causes also an erasure of the attractor, and finally a new attractor is constructed at another target object. The right image in Figure 4.4 shows the direct movement to the accompanying equilibrium of the new attractor.

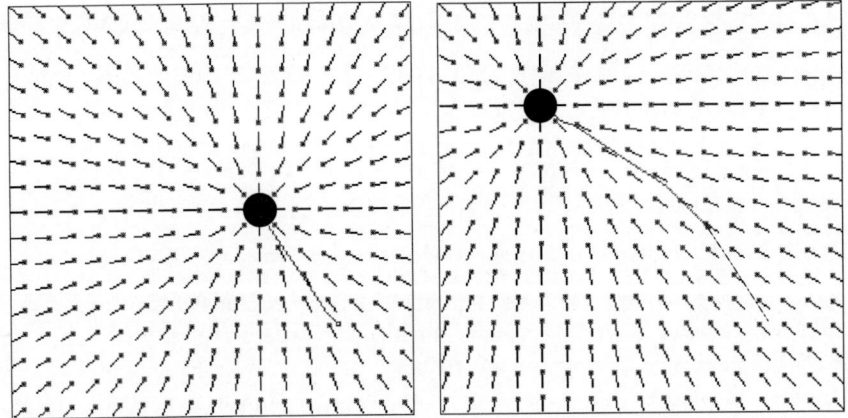

Fig. 4.4. (Left) Direct approaching the accompanying equilibrium of the first attractor; (Right) Direct approaching the accompanying equilibrium of the second attractor.

Another example is to visit a series of target objects with an *eye-on-hand robot arm* and visually inspect the objects in detail according to a certain succession. As soon as the inspection of an object is finished, *i.e.* a sub-task is completed, the attractor constructed at the object is transformed into a repellor, and additionally at the next object in the succession an attractor is specified. In consequence of this, the robot arm is repelled from the current object and attracted by the next one *(multiple goals achievement)*. Figure 4.5 shows in the left image the process of approaching the effector to the accompanying equilibrium of the first attractor (constructed at the first target object). Then, the object is inspected visually, and after completion the attractor is replaced by a repellor at that place, and finally a new attractor is constructed at another target object. The right image in Figure 4.5 shows the superposition of the attractor and the repellor vector field, and the course of effector movement which is repelled from the first object and attracted by the second one.

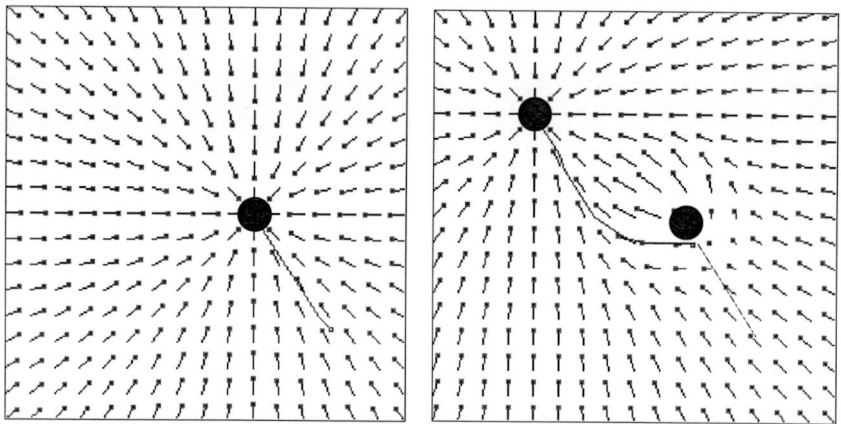

Fig. 4.5. (Left) Direct approaching the accompanying equilibrium of an attractor; (Right) Approaching the accompanying equilibrium of another attractor after being pushed away from the position, at which the former attractor changed to a repellor.

Dynamical Change of Current Deliberate Vector Fields

The dynamical construction of current deliberate vector fields from a generic deliberate vector field takes place during the process of task-solving. This on-line construction is necessary to keep the process under supervision by the relevant short-term plan of the overall plan. However, the task-solution can be reached only by real interaction with the environment which is done by visual feedback control as introduced above. First, the switch-off signal for completing a short-term plan and the trigger signal for starting a new short-term plan must be extracted from visual feedback of the environment. For example, in a task of robotic grasping one must recognize from the images a stable grasping pose, and this state will complete the elementary sub-task of approaching the target object, and the next elementary sub-task of closing the fingers can be executed. Second, in many sub-tasks the short-term plan should prescribe the course of effector movements only roughly, *i.e.* play the role of a supervisor, but the real, fine-tuned effector movements must be obtained by continual visual feedback. For example, in a task of manipulator navigation one must avoid obstacles which are supposed to be located on places slightly deviating from the plan. In this case, visual feedback control should be responsible to keep the manipulator at certain distances to the obstacles.

Reactive Vector Fields and Deliberate Vector Fields

The basic representation scheme for visual feedback control is a reactive vector field, and in the specific case of an elementary behavior it consists of just one attractor and nothing else (as discussed above). For this case, the current measurement vector in the image is transformed into a desired measurement vector iteratively. The desired measurement vector is the main feature for specifying the reactive vector field.

The question of interest is, how to combine the deliberate and the reactive part of the task-solving process in the methodology of vector fields.

For this purpose we regard the deliberate vector field more generally as a *memory* which is organized in two levels. The top level deals with the planning aspect and contains the centers of attractors, centers of repellors, and force vectors for reaching the equilibrium points. The bottom level includes the reactive aspect and contains a series of reactive vector fields which belong to certain attractors or repellors. For example, in the task of robotic grasping the attractor at the top level represents the final position of the robot hand roughly which is supposed to be relevant for grasping, and the reactive vector field at the bottom level is defined on the basis of a desired measurement vector which represents the pattern of a stable grasping situation. Therefore, the gripper can approach the target object based on the attractor in the top level, and can be fine-tuned on the basis of the relevant appearance pattern in the bottom level. Another example, in a task of manipulator navigation a repellor at the top level represents the rough position of an obstacle, and the desired measurement vector at the bottom level may represent the critical distance between manipulator and obstacle which should be surpassed.

Three-Layered Vertical Organization of Force Vector Fields

We can summarize that a *task-solving process is organized vertically* by three layers of force vector fields. These are the generic deliberate field at the top level, the layer of current deliberate fields at the middle level, and the current reactive fields at the bottom level. The deliberate field at the top level represents the overall task, the deliberate fields at the middle level describe the decomposition into elementary sub-tasks, and the reactive fields at the bottom level represent the actual task-solving process.[8] Previously, we distin-

---

[8] Müller proposes a three-layered model of autonomous agent systems [109] which fits to the vertical organization of our system. The top, middle, and bottom level is called *cooperation, planning, and reactive layer*, respectively.

guished between behaviors (elementary or assembled) and instructions which can be considered anew in the methodology of vector fields. An elementary behavior is based on a reactive vector field only, an assembled behavior is based on reactive vector fields and possibly includes deliberate vector fields, and an instruction is based on deliberate vector fields only. Generally, the actual effector movements are determined by considering a bias which can be represented in a deliberate field, and additionally considering visual feedback control which is represented in reactive fields.

However, for certain sub-tasks and environments it is reasonable to execute the relevant plan without frequent visual feedback. In these cases, the actual task-solving process is represented by the deliberate fields only. On the other hand, for certain sub-tasks visual feedback control must be applied without a deliberate plan. After completing a sub-task of this kind the covered course or the final value of variable state vector of the effector could be of interest for successive sub-tasks. Therefore, a possible intention behind a behavior is to extract grounded informations by interaction with the environment. Especially, the grounded informations can contribute to the generation or modification of deliberate fields, *i.e. biasing successive sub-tasks*. Related to the organization of vector fields, we can conclude that in general a *bidirectional flow of information* will take place, *i.e.* from top to bottom and/or reverse.

Horizontal Organization of Force Vector Fields for Various Effectors

So far, it has been assumed that a high-level, deliberate task is treated by actively controlling just one effector of a camera-equipped robot system. Only one generic deliberate field was taken into account along with the offspring fields at the middle and bottom level. However, in general a high-level robotic task must be solved by a camera-equipped robot system with more than one active effector. *Several effectors must contribute* for solving a high-level, deliberate task and maybe even for solving a sub-task thereof. Previously, we stipulated that in each elementary sub-task the camera-equipped robot system is trying to reach just one deliberate *goal* and keep certain *constraints*. The critical issue concerns the complexity of the goal and/or the constraints, because, based on this, the number and types of simultaneously active effectors are determined. For example, a surveillance task may consist of approaching a vehicle to a target position and simultaneously rotating a mounted head-camera for fixating an object at which the vehicle is passing by. In this case, two effectors must work *simultaneously* in a synchronized mode, *i.e.* the vehicle position and the head orientation are interrelated. Apart from the simultaneous activity of several effectors it is usual that several effectors come into play one after the other in a *sequential* mode.

Splitting the Variable State Vector of Effectors

Regardless of the simultaneous or the sequential mode, for each effector at least one generic deliberate field is required along with the offspring fields. However, for certain effectors it makes sense to split up the variable state vector $S^v(t)$ into sub-vectors $S^{v1}(t), \cdots, S^{vn}(t)$. Corresponding to this, also the control vector $C(t)$ must be split up into sub-vectors $C^{v1}(t), \cdots, C^{vn}(t)$. As a first example, the vector of pose parameters of a robot hand can be split up in the sub-vector of position parameters (*i.e.* 3D coordinates X, Y, Z) and the sub-vector of orientation parameters (*i.e.* Euler angles *yaw, pitch, roll*). As a second example, the view parameters of stereo head-cameras can be treated as single-parameter sub-vectors which consist of the pan, the tilt, and the vergence angles. The splitting up of the variable state vector depends on the type of the effector and the type of the task. Furthermore, in certain tasks it is reasonable to select a certain sub-vector of parameters as being variable, and keep the other sub-vector of parameters constant. In conclusion, in the designing phase we must specify for each effector of the camera-equipped robot system a set of sub-vectors of parameters which will be potentially controlled for the purpose of solving the underlying sub-task. For each effector this set can be empty (*i.e.* the effector is be kept stable), or the set may contain one sub-vector (*i.e.* the effector can perform just one category of movements), or the set may contain several sub-vector (*i.e.* the effector can perform several categories of movements).

Vertical and Horizontal Organization of Force Vector Fields

Generally, for each effector a set of generic deliberate fields is required together with the offspring fields, which depends on the complexity of the high-level task. The number of generic deliberate fields corresponds with the number of sub-vectors of variable parameters. In consequence of this, apart from the vertical organization (discussed previously) the *task-solving process* must also be organized *horizontally* including more than one generic deliberate fields. Figure 4.6 shows figuratively the vertical and horizontal organization of deliberate and reactive fields, which may be involved in a task-solving process for a high-level, deliberate task. In the case that image-based effector servoing is applied without a deliberate plan, then the deliberate fields are trivial. In the case that a deliberate plan is executed without frequent visual feedback, then the reactive fields are trivial. Information is exchanged *vertically and horizontally* which is indicated in the figure by bidirectional arrows.

Monitoring the Task-Solving Process

In the designing phase of a task-solving process the high-level, deliberate task must be decomposed into sub-tasks and for these one must regulate the *way of cooperation*. We distinguish two modes of cooperation, *i.e.* the

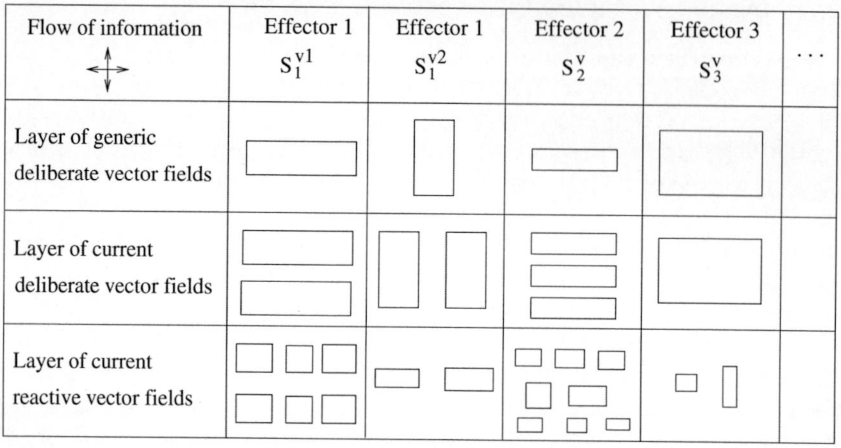

Flow of information ←+→ ↓	Effector 1 S_1^{v1}	Effector 1 S_1^{v2}	Effector 2 S_2^{v}	Effector 3 S_3^{v}	...
Layer of generic deliberate vector fields					
Layer of current deliberate vector fields					
Layer of current reactive vector fields					

Fig. 4.6. Vertical and horizontal organization of deliberate and reactive fields involved in a task-solving process.

sub-tasks are treated *sequential or simultaneous* with other sub-tasks. In the sequential mode several goals must be reached, but one after the other. The completion of a certain sub-task contributes current information or arranges an environmental situation which is needed for successive sub-tasks. Related to the methodology of *dynamical systems* these contributions are reflected by changing certain vector fields. For example, after grasping, picking-up, and removing a target object we have to make topical the relevant deliberate vector field, *i.e.* removing the attractor which has been constructed at the object. In the simultaneous mode of cooperating sub-tasks, several goals are pursued in parallel. Maybe, several sub-tasks are independent and can be treated by different effectors at the same time for the simple reason of saving overall execution time. On the other hand, maybe it is mandatory that several sub-tasks cooperate in a synchronized mode for solving a high-level task.

Three Categories of Monitors

A so-called *monitor* is responsible for supervising the task-solving process including the cooperation of sub-tasks.[9] More concretely, the monitor must take care for three aspects. First, for each sub-task there is a limited period of time, and the monitor must *watch a clock* and wait for the signal which indicates the finishing of the sub-task. If this signal is coming timely, then the old sub-task is switched off and the successive one switched on. However, if this signal is coming belated or not at all, then exception handling is

---

[9] Kruse and Wahl [91] presented a camera-based monitoring system for mobile robot guidance. We will use the term *monitor* more generally, *e.g.* including tasks of keeping time limitations, confirming intermediate situations, and treating exceptional situations.

needed (*e.g.* stopping the overall process). Second, each sub-task should contribute topical information or arrange an environmental situation, and after completion of the sub-task the monitor must check whether the *relevant contribution* is supplied. For example, topical information maybe is collected in deliberate fields, and an analysis of the contents is required to decide whether the successive sub-task can be started. Third, any technical system, which is embedded in a real environment, must deal with *unexpected events*. For example, maybe a human is entering the working area of certain effectors illegally. The monitor must detect these events and react appropriately, *e.g.* continue after a waiting phase or final stopping the process. The monitors are implemented in the designing phase and are specific for each high-level task.

Cooperation of Sub-tasks

One must determine which sub-tasks should work sequentially or simultaneously and which period of time is supposed to be acceptable. Furthermore, for each sub-task the constituents of the environment must be determined which are normally involved. A constituent of the environment can be the actuator system (*e.g.* diverse effectors), the cameras (mobil or stable), or several task-relevant objects (*e.g.* stable platform), *etc.* These fundamental informations (concerning periods of time or constituents of the environment) are needed for checking relevant contributions or detecting events during the task-solving process. Figure 4.7 shows for a task-solving process the cooperative arrangement of sub-tasks, *i.e.* sequentially or simultaneously, together with the environmental constituents involved in each sub-task (indicated by a dot). It is just a generic scheme which must be made contrete for each specific task during the designing phase.

4.2.2 Generic Modules for System Development

For supporting the designing phase of autonomous camera-equipped robot systems we discovered 12 *generic modules*. They are generic in the sense that *task-specific modules* will make use of them with specific parametrizations or specific implementations.[10] Each generic module is responsible for solving an elementary sub-task. Regarding the explanations in the previous subsection we distinguish *three instructional modules, six behavioral modules,* and *three monitoring modules*. The instructional modules are based on deliberate vector fields, the behavioral modules generate reactive vector fields and possibly are based on deliberate vector fields. The monitoring modules play an exceptional

---

[10] The generic modules serve as *design abstractions* in order to simplify system development for a specific robotic task. This methodology is similar to the use of general *design patterns* for the development of object-oriented software products [61]. However, we propose application-specific design patterns for the development of autonomous, camera-equipped robot systems.

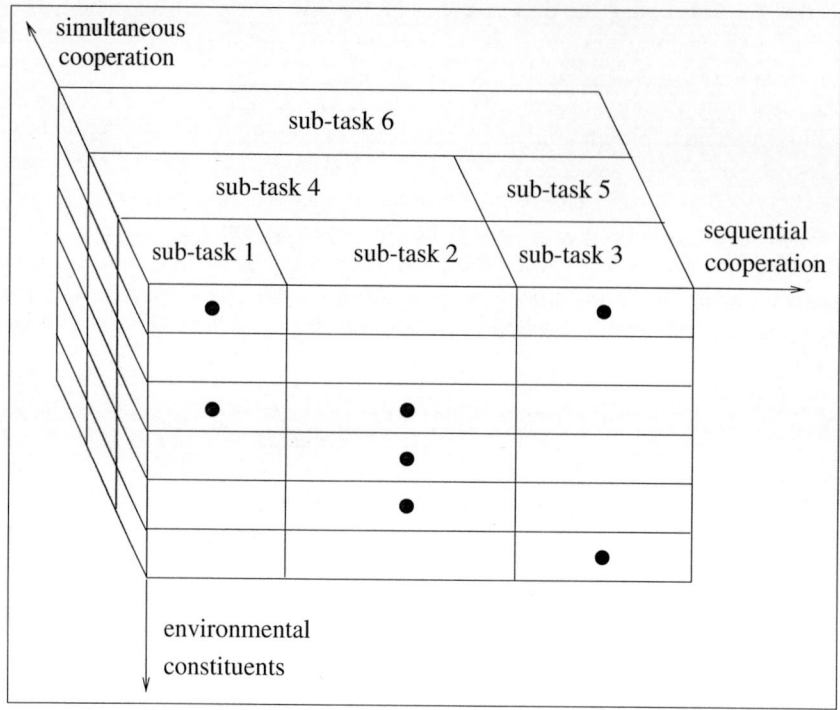

Fig. 4.7. Sequential and simultaneous cooperation of sub-tasks and involved environmental constituents.

role, *i.e.* they do not generate and are not based on vector fields. In Section 4.3 we will show the usage of a subset of 9 generic modules (three instructional, behavioral, and monitoring modules) for an exemplary task. Therefore, this subsection presents just these relevant modules, and the remaining subset of 3 behavioral modules is presented in Appendix 2.

Generic Module MI_1 for Assembled Instruction

The module involves the human designer of the autonomous system. With the use of the control panel a certain effector can be steered step by step into several states. The relevant succession of elementary instructions is obtained from a deliberate field which may contain a trajectory. Based on this *teach-in approach*, the designer associates proprioceptive states with certain exteroceptive features, *e.g.* determines the position of a static environmental object in the basis coordinate system of the robot arm. The list of teached states is memorized. No explicit, deliberate field is involved with this module, except the one in the mind of the human operator.

Module MI_1

1. Determine relevant type of variable state vector ≀S^v≀.
2. Manually generate a state vector $S^v(t)$ of the relevant effector.
3. Memorize pair $(t, S^v(t))$ of demonstration index and new state vector,
   and increment demonstration index t := t + 1.
4. If (demonstration_index $\leq$ required_number_of_demonstrations) then go to 2.
5. Stop.

Generic Module MI_2 for Assembled Instruction

The module represents a steering mechanism which changes the variable state of an effector step by step according to a *pre-specified trajectory*. The relevant succession of elementary instructions is obtained from a deliberate field which may contain a trajectory. The purpose is to bring the effector into a certain state from which to start a successive sub-task. The concrete course has been determined in the designing phase. No visual feedback control is involved, however the movement is organized incrementally such that a monitor module can interrupt in exceptional cases (see below monitor MM_3).

Module MI_2

1. Determine relevant type of variable state vectors ≀S^v≀.
   Take deliberate field of ≀S^v≀ into account.
2. Determine current state vector $S^v(t)$ of the relevant effector.
3. Determine control vector according to equation
   $C(t) := VF_O[trajectory\_for\_S^v](S^v(t))$.
4. If ($\| C(t) \| \leq \eta_1$) then go to 7.
5. Change variable state vector according to equation
   $S^v(t+1) := f^{ts}(C(t), S^v(t))$,
   and increment time parameter t := t + 1.
6. Go to 2.
7. Memorize final state vector $S^v(t)$, and stop.

Generic Module MI_3 for Assembled Instruction

The module is similar to the previous one in that the relevant succession of elementary instructions is obtained from a deliberate field. Additionally, image measurements are taken during the step-wise change of the effector state. The measurements are memorized together with the time indices, and the states of the effector. The purpose is to *collect image data, e.g.* for learning coordinations between effector state and image measurements, for learning operators for object recognition, or for inspecting large scenes with cameras of small fields of view.

Module MI_3

---

1. Determine relevant type of variable state vectors $\wr S^v \wr$ and accompanying type of measurements $\wr Q \wr$.
   Take deliberate field of $\wr S^v \wr$ into account.
2. Determine current state vector $S^v(t)$ of the relevant effector.
3. Determine control vector according to equation
   $C(t) := VF_O[trajectory\_for\_S^v](S^v(t)).$
4. If ($\| C(t) \| \leq \eta_1$) then go to 9.
5. Change variable state vector according to equation
   $S^v(t+1) := f^{ts}(C(t), S^v(t)),$
   and increment time parameter $t := t + 1$.
6. Determine new measurement vector $Q(t)$.
7. Memorize triple $(t, S^v(t), Q(t))$ of time index, new state vector, and new measurement vector.
8. Go to 2.
9. Stop.

In the following we present three behavioral modules which will be used in the next section. Further behavioral modules are given in Appendix 2.

Generic Module MB_1 for Elementary Behavior

The module represents a *visual feedback control* algorithm for the variable state vector of a robot effector. The current measurements in images should change step by step into desired measurements. The control function is based on approximations of the relationship between changes of the effector state and changes of measurements in the images, *e.g.* linear approximations with Jacobian matrices. Apart from the mentioned relationship no other bias is included, *e.g.* no plan in form of a deliberate field. The generated effector trajectory is a pure, reactive vector field.

Module MB_1

---

1. Take desired measurement vector Q^* into account.
2. Determine current measurement vector $Q(t)$.
3. Determine current state vector $S^v(t)$ of the relevant effector.
4. Determine control vector according to equation
   $C(t) := f^{ct}(Q^*, Q(t), S^v(t))$.
5. If ($\| C(t) \| \leq \eta_2$) then go to 8.
6. Change variable state vector according to equation
   $S^v(t+1) := f^{ts}(C(t), S^v(t))$,
   and increment time parameter t := t + 1.
7. Go to 1.
8. Return final state vector $S^v(t)$, and stop.

Generic Module MB_2 for Assembled Behavior

The module is responsible for an assembled behavior which integrates two elementary behaviors, *i.e.* executing *two goal-oriented cycles*. While trying to keep desired image measurements of a first type, the robot effector keeps on changing its variable state to reach desired image measurements of a second type. The inner cycle takes into account the first type of measurements and gradually changes the effector state such that the relevant desired measurement is reached. Then, in the outer cycle one changes the effector state a certain extent such that the image measurement of the second type will come closer to the relevant desired measurement. This procedure is repeated until the desired measurement of the second type is reached. No plan in form of a deliberate field is used, and the generated effector trajectory is a pure, reactive vector field. However, the resulting course of the effector state is memorized, *i.e.* will be represented in deliberate fields.

Module MB_2

---

1. Determine relevant type of variable state vectors $\wr S^{v1} \wr$ and accompanying type of measurements $\wr Q^{v1} \wr$.
   Initialization of a deliberate field for $\wr S^{v1} \wr$.
   Determine relevant type of variable state vectors $\wr S^{v2} \wr$ and accompanying type of measurements $\wr Q^{v2} \wr$.
   Initialization of a deliberate field for $\wr S^{v2} \wr$.
2. Behavioral module MB_1:
   Configure with ($\wr S^{v1} \wr$, $\wr Q^{v1} \wr$), execution, and return ($S^{v1}(t)$).
3. Construct an equilibrium in the deliberate field $\wr S^{v1} \wr$ based on current state vector $S^{v1}(t)$.
4. Take desired measurement vector $(Q^{v2})^*$ into account.
5. Determine current measurement vector $Q^{v2}(t)$.
6. Determine current state vector $S^{v2}(t)$ of the relevant effector.
7. Determine control vector according to equation
   $C^{v2}(t) := f^{ct}((Q^{v2})^*, Q^{v2}(t), S^{v2}(t))$.
8. If ($\| C^{v2}(t) \| \leq \eta_2$) then go to 12.
9. Change variable state vector according to equation
   $S^{v2}(t+1) := f^{ts}(C^{v2}(t), S^{v2}(t))$, and increment time parameter t $:= t + 1$.
10. Construct an equilibrium in the deliberate field $\wr S^{v2} \wr$ based on new state vector $S^{v2}(t)$.
11. Go to 2.
12. Memorize final deliberate fields $\wr S^{v1} \wr$ and $\wr S^{v2} \wr$, and stop.

Generic Module MB_3 for Assembled Behavior

The module represents a steering mechanism which changes the variable state of an effector step by step according to a pre-specified course. The relevant succession of elementary instructions is obtained from a deliberate field which may contain a trajectory. In distinction to the instructional module MI_2 certain measurements are taken from the image continually, and if certain *conditions hold*, then the plan execution is *interrupted*. At the bottom-level (of our vertical organization) no reactive vectors are generated, because visual feedback only serves for recognizing the stopping condition.

Module MB_3

---

1. Determine relevant type of variable state vectors $\wr S^v \wr$.
   Take deliberate field of $\wr S^v \wr$ into account.
   Determine relevant type of measurements $\wr Q \wr$.
2. Take desired measurement vector Q^* into account.
3. Determine current measurement vector $Q(t)$.
4. If ($\parallel Q^* - Q(t) \parallel \leq \eta_2$) then go to 10.
5. Determine current state vector $S^v(t)$ of the relevant effector.
6. Determine control vector according to equation
   $C(t) := VF_O[trajectory\_for\_S^v](S^v(t))$.
7. If $\parallel C(t) \parallel \leq \eta_1$ then go to 10.
8. Change variable state vector according to equation
   $S^v(t+1) := f^{ts}(C(t), S^v(t))$,
   and increment time parameter t := t + 1.
9. Go to 2.
10. Memorize final state vector $S^v(t)$, and stop.

The following three monitor modules are responsible for the surveillance of the task-solving process. Three important aspects have been mentioned above and three generic modules are presented accordingly.

Generic Module MM_1 for Time Monitor

The *time monitor* MM_1 is checking whether the sub-tasks are solved timely (see above for detailed description).

Module MM_1

---

1. Take period of time for sub-task i into account.
2. Determine working state of sub-task i.
3. If (sub-task_i_is_no_more_working) then go to 7.
4. Determine current time.
5. If (sub-task_i_is_still_working_and_current_time_within_period)
   then wait a little bit, then go to 2.
6. If (sub-task_i_is_still_working_and_current_time_not_within_period)
   then emergency exit.
7. Stop.

Generic Module MM_2 for Situation Monitor

The *situation monitor* MM_2 is checking after completion of a certain sub-task, whether an environmental situation has been arranged which is needed in successive sub-tasks (see above for detailed description).

Module MM_2

---

1. Determine working state of sub-task i.
2. If (sub-task_i_is_still_working) then wait a little bit then go to 1.
3. Take goal situation of sub-task i into account.
4. Take actual situation after completing sub-task i into account.
5. Determine a distance measure between goal and actual situation.
6. If (distance_measure_is_beyond_a_threshold) then emergency exit.
7. Stop.

Generic Module MM_3 for Exception Monitor

The *exception monitor* MM_3 observes the overall task-solving process with the purpose of reacting appropriately in case of unexpected events (see above for detailed description).

Module MM_3

---

1. Take overall period of time for the high-level task into account.
2. Throughout the period of time, for all sub-tasks i:
2.1. Determine working state of sub-task i.
2.2. Determine working space of sub-task i.
2.3. If (sub-task_i_is_working_and_unexpected_event_in_
    relevant_working_space) then emergency exit.
3. Stop.

Scheme of Task-Specific Modules MT_i

Specific implementations of the basic, generic modules will be used in task-specific modules which are responsible for solving certain tasks and sub-tasks. For simplifying the designing phase we will configure *task-specific modules* at several levels of an abstraction hierarchy, such that in general higher-level modules are based on combinations of lower-level modules. The bottom level of the abstraction hierarchy contains specific implementations of the generic modules introduced above.

Generally, the scheme of a task-specific module consists of three components, *i.e.* combination of lower-level modules, functions for pre-processing and post-processing relevant data, and an input/output mechanism for reading from and writing to the memory. Concerning the combination of lower-level modules we distinguish between *sequential* (denoted by &) and *parallel* execution (denoted by |). The parallel combination is asynchronous in the sense that the control cycles for two participating lower-level modules will not be synchronized between each other. Instead, synchronization takes place in form of simultaneously starting some lower-level modules and waiting until all of them have finished, and then starting the next module(s) in the succession. However, if synchronization would be needed at a more fine-grained level of control, then it must be implemented as a basic module (*e.g.* see above the assembled behavior MB_2). As a result of the lower-level modules one obtains topical image *information about the scene*, and/or topical *states of the effectors*.

For applying lower-level modules, perhaps certain functions for extracting relevant images features must be applied in advance. For parameterizing the functions specifically one may obtain data from local input of the current task-specific module, *i.e.* parameter list of task-specific module, or from global input, *i.e.* shared memory of the task-solving process. Alternatively, a function may implement an iterative, feedback-based approach of feature extraction in which the parameters are tuned autonomously.[11] The results of the pre-processing functions are forwarded to the lower-level modules, and the results of these modules are forwarded to the output (maybe after applying post-processing functions). Local input or output contains intermediate data and global input or output contains final data which are represented in the shared memory for solving the high-level task.

In summary, the *generic scheme of task-specific modules* contains the following entries.

Task-specific module MT_i

1. Name
2. Local input
3. Global input
4. Pre-processing functions
5. Combination of lower-level modules
6. Post-processing functions
7. Local output
8. Global output

In the next Section 4.3 we present some examples of task-specific modules. However, only the entry *combination of lower-level modules* will be specified in detail and the others will remain vague.

---

[11] Feedback-based autonomous image analysis is one of the characteristics of Robot Vision (see Subsection 1.2.2).

4.3 Treatment of an Exemplary High-Level Task

In Section 4.1 we explained that a bottom-up designing methodology is essential for obtaining an autonomous camera-equipped robot system. The application phase of a task-solving process must be preceded by an experimentation phase whose outcome is supposed to be a configuration of modules for solving the underlying task. In the preceding Section 4.2 we presented basis mechanisms and categories of generic modules which must be implemented specifically. This section describes an exemplary high-level task, applies the bottom-up designing methodology, and presents specific implementations of the generic modules.

4.3.1 Description of an Exemplary High-Level Task

The high-level deliberate task is to find a target object among a set of objects and carry it to another place for the purpose of detailed inspection. Figure 4.8 shows the original scene including the robot system and other task-relevant environmental constituents.

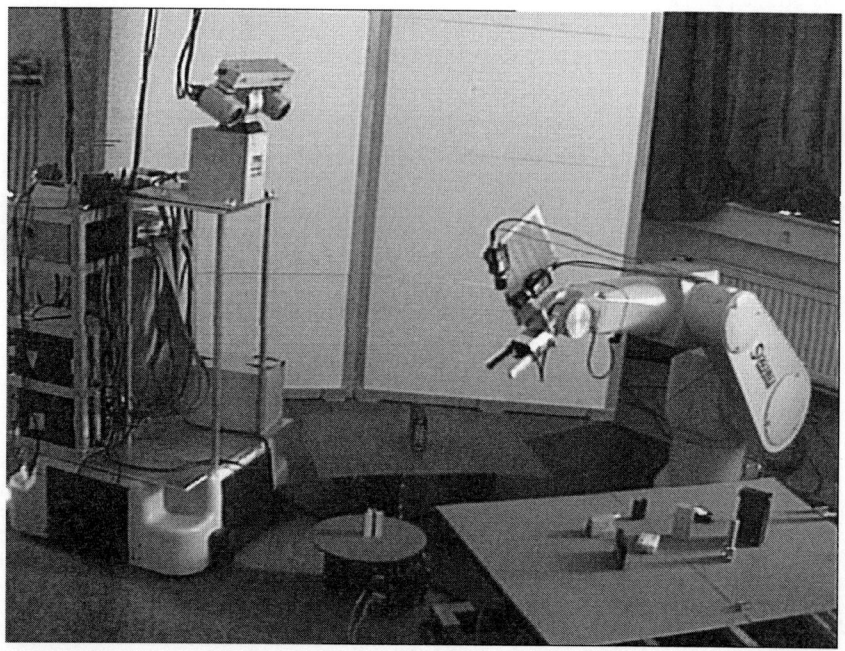

Fig. 4.8. The image shows the scene including robot arm, robot vehicle with binocular head, rotary table, and the task-relevant domestic area, inspection area, and parking area.

The computer system consists of a Sun Enterprise (E4000 with 4 Ultra-Sparc processors) for doing image processing and of special purpose processors for computing the inverse kinematics and motor signals. The actuator system of the robot system is composed of four subsystems. The first subsystem is a *robot arm* (Stäubli–Unimation RX–90) fastened on the ground plane. Based on six rotational joints one can move the robot hand in arbitrary position and orientation within a certain working space. Additionally, there is a linear joint at the robot hand for opening/closing parallel jaw fingers. The position of the robot hand is defined by the tool center point, which is fixed in the middle point between the two finger tips (convenient for our application). The second subsystem is a *robot vehicle* (TRC Labmate) which can move arbitrary in the neighborhood of the robot arm, *e.g.* turn round and translate in any direction. The position of the robot vehicle is defined as the center of the platform. The third subsystem is a *robot head* including a stereo camera (TRC bisight) which is fastened on the robot vehicle. The robot head is equipped with pan, tilt, and vergence degrees-of-freedom (DOF), and zooming/focusing facilities. By moving the vehicle and changing view direction of the cameras one can observe the robot arm and its working space under different viewing points. The fourth subsystem is a *rotary table* on which to place and to rotate objects in order to inspect them from any view.

Further constituents of the scene are three particular ground planes which must be located within the working space of the robot hand (of the robot arm). These task-relevant planes are the so-called *domestic area*, the *inspection area*, and the *parking area*. We assume, that several objects can be found on the domestic area. A specific target object is of further interest and should be inspected in detail which is done at the inspection area. For this purpose, the target object must be localized on the domestic area, carried away, and placed on the inspection area. However, it may happen that the target object can not be approached for robotic grasping due to obstacle objects. In this case, the obstacle objects must be localized, moved to the parking area and placed there temporary.

Five Cameras for the Robot System

The autonomy in solving this task is supposed to be reached with *five cameras*. First, for the purpose of surveillance of the task-solving process one camera (so-called *ceiling-camera* CA_1) is fastened at the ceiling. The optical axis is oriented into the center of the working area, and the objective is of middle focal length (*e.g.* 6 mm) such that the whole scene is contained in the field of view. Second, two cameras are fastened at the robot hand, the first one (so-called *hand-camera* CA_2) is used for localizing the target object and the second one (so-called *hand-camera* CA_3) is used for controlling the grasping process. For hand-camera CA_2 the viewing direction is approximately parallel to the fingers and the objective is of small focal length (*e.g.*

$4.2mm$), *i.e.* a large part of the domestic area should be observable at a reasonable resolution. For hand-camera CA_3 the viewing direction is straight through the fingers (*i.e.* approximately orthogonal to the one of CA_2), and the objective is of middle focal length (*e.g.* $6mm$) such that both the fingers and a grasping object are contained in the field of view. Third, two cameras of the robot head (so-called *head-cameras* CA_4 and CA_5) are used for controlling the placement of obstacle objects at the parking area or of the target object at the inspection area. The head-cameras are used also for detailed object inspection. Depending on the category of the carried object (obstacle or target object) the DOFs of the robot head must be changed appropriately such that the parking area or the inpection area will appear in the field of view, respectively. The pan and tilt DOF of the robot head are from $-90°$ to $+90°$ degrees each. The vergence DOF for each camera is from $-45°$ to $+45°$ degrees. The focal length of the head-cameras can vary between $11mm$ and $69mm$.

Abstract Visualization of the Original Scene

Figure 4.9 shows an abstract depiction of the original scene. For the purpose of task decomposition we introduce *five virtual points* P_1, P_2, P_3, P_4, P_5, which will serve as intermediate positions of the trajectories of the robot hand (see below). Generally, these positions are represented in the coordinate system of the robot arm which is attached at the static basis. Position P_1 is the starting point of the robot hand for solving the task, positions P_2 and P_3 are located near the domestic area and serve as starting points for actively locating the target object and grasping the objects, respectively, position P_4 is the point from which to start the servoing procedure for placing an obstacle object on the parking area, and finally position P_5 is the point from which to start the servoing procedure for placing the target object on the inspection area.

The decomposition of the high-level, deliberate task into sub-tasks, and the configuration and implementation of *task-specific modules* is based on an experimental designing phase. Therefore, in the following subsections the designing phase and the application phase are explained in coherence for each sub-task.

4.3.2 Localization of a Target Object in the Image

In the application phase the *first goal* is to find a target object among a set of objects which are located on the domestic area.

Designing Aspects for Localization of a Target Object

For the localization of a target object an operator is needed which should be robust and efficient (as has been studied in Chapters 2 and 3, extensively). Usually, the application phase leaves open certain *degrees of freedom in arranging cameras* and taking images. It is an issue of the designing phase to

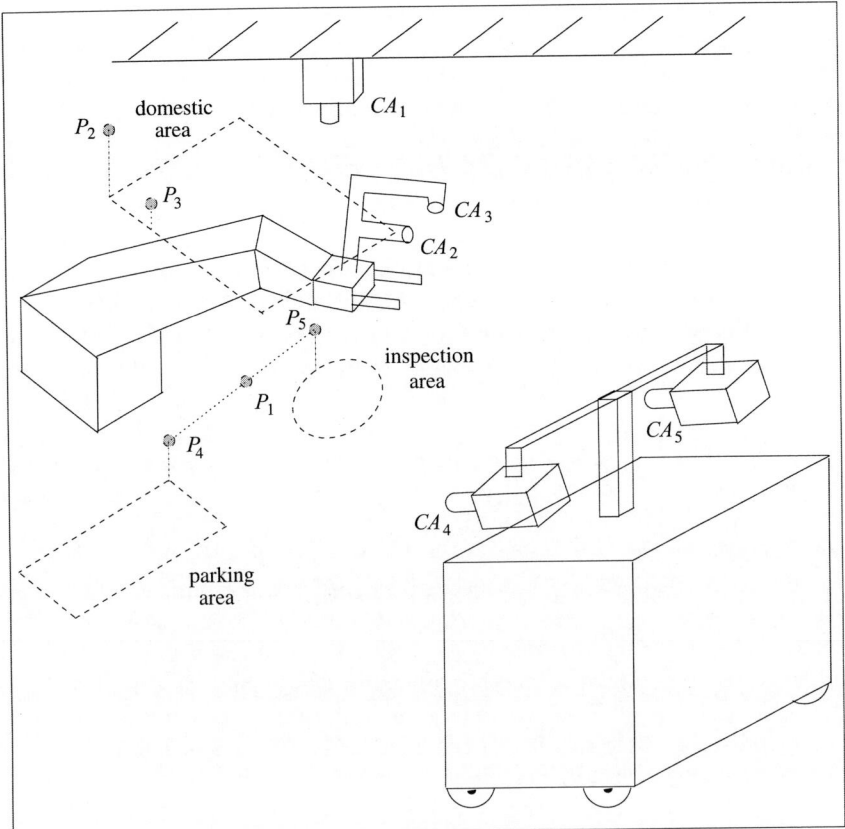

Fig. 4.9. Abstract visualization of the original scene.

determine viewing conditions which reveal optimal robustness and efficiency in object localization. A prerequisite for obtaining robustness is to keep *similar viewing conditions* during the learning phase and the application phase. In general, it is favourable to take top views from the objects instead of taking side views because in the latter case we have to deal with occlusions. However, there may be constraints in taking images, *e.g.* limited free space above the collection of objects. These aspects of the application phase must be considered for arranging similar conditions during the learning phase. A prerequisite for obtaining efficiency is to keep the complexity of the appearance manifold of a target object as low as possible. This can be reached in the application phase by constraining the possible relationships between camera and target object and thus reducing the variety of viewing conditions.

In the designing phase, we demonstrate views from the target object and from counter objects (for the purpose of learning thereof), but consider constraints and ground truths and exploit degrees of freedom (which are supposed to be relevant in the application phase). Based on this, the system learns operators as robust and efficient as possible (under the supervision of the designer). Additionally, the system should come up with desirable geometric relationships between camera and object (subject to the degrees of freedom in the application phase). For example, as a result of the learning process we may conclude that *objects should be observed from top* at a certain distance to the ground and the optical axis of the hand-camera CA_2 should be kept normal to the ground plane. In Figure 4.9 we introduced a virtual point P_2, which is straight above a certain corner of the domestic area and the normal distance from this area is the optimal viewing distance.

Additionally, we may conclude that under this optimal viewing distance the field of view of the camera is less than the size of the domestic area. However, the set of objects is spread throughout the whole area and only a sub-set can be captured in a single image. Consequently, in the application phase the *robot arm has to move the hand-camera CA_2* horizontally over the domestic area (at the optimal viewing distance, which defines the so-called *viewing plane*) and take several images step by step. For example, a horizontal meander-type movement would be appropriate with the starting point P_2. Figure 4.10 shows an intermediate step of this movement. On the left, the domestic area is shown and the hand-cameras, the optical axis of the camera CA_2 is directed normal (approximately) to the ground. The image on the right is obtained by the hand-camera CA_2 which depicts only a part of the domestic area. Based on the specific interplay between the characteristics of sub-task, environment, and camera, it is necessary to execute *camera movements* according to a certain strategy.

Fig. 4.10. (Left) Domestic area and hand-cameras; (Right) Part of the domestic area taken by the hand-camera.

The specific shape of the meander-type movement must be determined such that a complete image can be constructed from the large domestic area, *i.e.* the complete image is fitted together from single images. Generally, the stopping places of the hand-camera CA_2 should be chosen such that the collection of single images does not leave any holes in the domestic area which may not be captured. On the other hand, one should avoid that single images in the neighborhood do capture too much overlap, because this reduces efficiency in object localization due to repeated applications in the overlapping image areas. Interestingly, our approach of object recognition requires a certain degree of overlap which is based on the size of the rectangular object pattern used for recognition. In the case that the target object is partly outside the field of view, the learned operator for target localization can not be applied successfully to this image. However, if we arrange an overlap between neighbored images of at least the expected size of the target pattern, then the object is fully contained in the next image and can be localized successfully.

In consequence of this, for determining the relevant increments for the *stopping places of the hand-camera* CA_2 (and taking images) we must take into account the size of the *appearance pattern of the target object*.[12] Furthermore, a kind of calibration is needed which determines the relationship of pixel number per millimeter (see Subsection 4.4.2 later on). The principle is demonstrated in Figure 4.11. On top left the size of the image is depicted and on top right the size of the rectangular object pattern. On bottom left and right the domestic area is shown (bold outlined), with the meander and stopping places of the hand-camera CA_2 depicted on the left, and the series of overlapping images depicted on the right. The size of the rectangular object pattern correlates with the overlap between neighbored single images, and consequently the target pattern is fully contained in a certain image regardless of its location on the domestic area (*e.g.*, see the two occurrences).

Generally, the locations of all three particular ground planes (domestic area, parking area, and inspection area) are determined by a *teach-in* approach. The designer uses the control panel of the robot arm and steers the tip of the robot hand in succession to certain points of the particular areas, *e.g.* four corner points of the rectangular domestic and parking area, respectively, and a point at the center and the boundary of the circular inspection area, respectively. The respective positions are determined and memorized automatically based on inverse kinematics. In consequence of this, the positions of these areas are represented in the basis coordinate system of the robot arm. The virtual positions P_1, P_2, P_3, P_4, P_5 are also represented in this coordinate system and the specific relations to the three areas are based on certain experiments.

Position P_1 is specified dependent on the relationship between parking and inspection area (*e.g.* middle point between the areas). Position P_2 is specified based on the location of the domestic area. Specifically, as a result of experiments on learning operators for object recognition one determines the relation between P_2 and the domestic area, as explained above. Starting at position P_1 the robot hand must move to position P_2 and there continue

---

[12] This strategy shows exemplary the degree by which robot effector movements and image analysis techniques must work together, *i.e.* perception and action are strongly correlated for solving sub-tasks.

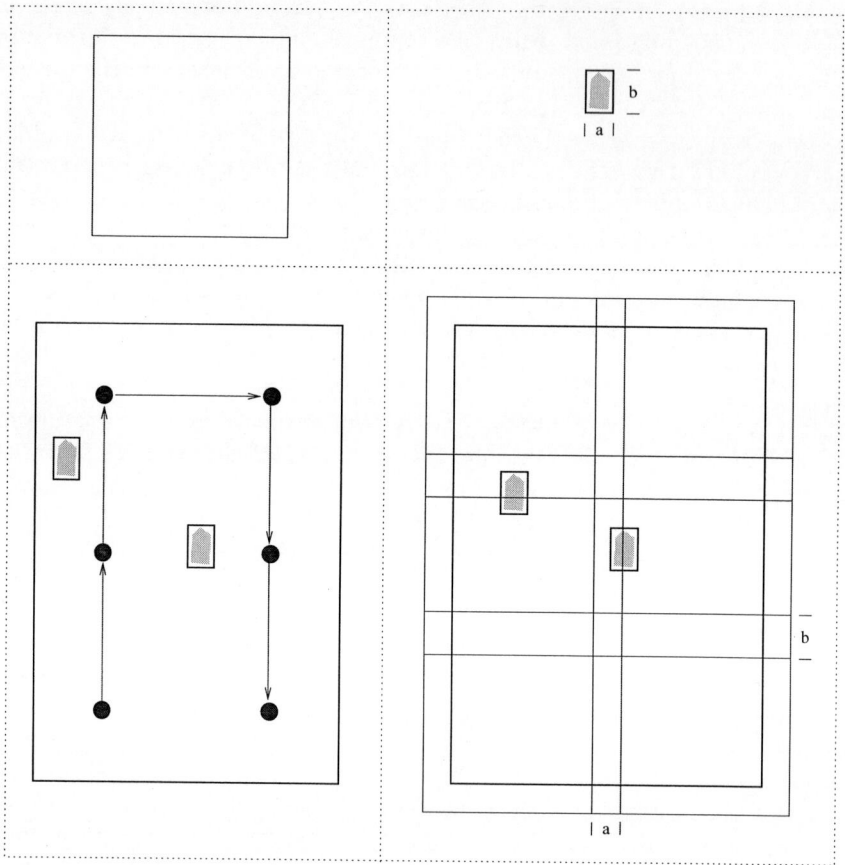

Fig. 4.11. Meander-type movement of hand-camera CA_2 and taking images with an overlap which correlates with the size of the rectangular object pattern.

with a meander-type movement over the domestic area. The movement of the robot hand from position P_1 to P_2 may also be specified by a teach-in approach, *i.e.* the designer supplies intermediate positions for approximating the desired trajectory. The shape and stopping places of the meander-type movement over the domestic area are determined based on experiments on object localization.

Task-Specific Modules for Localization of a Target Object

As a result of the experimental designing phase we define a task-specific module MT_1 which is based on the execution of a generic module of type MI_1, followed by a generic module of type MI_2, followed by a generic module of type MI_3, followed again by a generic module of type MI_2.

$$MT_1 \quad := \quad (MI_1 \quad \& \quad MI_2 \quad \& \quad MI_3 \quad \& \quad MI_2) \qquad (4.8)$$

The module MI_1 is expecting from a human operator to steer the robot hand to certain points on the domestic area in order to obtain relevant positions in the coordinate system of the robot arm. Especially, also the starting position P_2 of the meander-type movement over the domestic area can be determined thereof. Furthermore, the human operator must teach the system a certain trajectory from original position P_1 to position P_2. The first occurrence of module MI_2 is responsible for moving the robot hand from position P_1 to position P_2 along the specified trajectory. The module MI_3 is responsible for the meander-type movement over the domestic area and taking a series of images. The second occurrence of module MI_2 moves the robot hand back to starting position P_2. Pre-processing functions must determine the meander structure on the viewing plane over the whole domestic area, and define (few) intermediate stops as discussed above. Post-processing functions will localize the target object in the collection of single images. The output of the module is an index of the image containing the target object and relevant position in the image.

The generic module MI_3 (explained in Subsection 4.2.2) has been applied for taking images during the application phase. In addition to this, this generic module can also be applied during the designing phase for two other purposes. First, we must take images from the target and from counter objects under different viewing conditions such that the system can *learn an operator for object recognition* (see Section 3.2). Second, we must take images from artificial or natural calibration objects such that the system can approximate the transformation between *image and robot coordinate systems* (see Subsections 4.4.1 and 4.4.2 later on). The responsible task-specific modules are similar to MT_1 and therefore are not presented in this work.

4.3.3 Determining and Reconstructing Obstacle Objects

In the application phase the *second goal* is to determine and reconstruct possible obstacle objects which prevent the robot hand from approaching the target object.

Designing Aspects for Determining/Reconstructing Obstacles

In order to design task-specific modules for determining obstacle objects we must deal with the following aspects.

The purpose of reaching the target object is to grasp it finally. The critical issue of object grasping is to arrange a stable grasping situation under the constraint of occupying little space in the grasping environment (the latter is for simplifying the problem of obstacle avoidance during grasping). The hand of our robot arm is equipped with parallel jaw fingers, and additional space is occupied by the two hand-cameras and the fixing gadget (see Figure 4.10 (left)). According to the specific architecture it is favourable to grasp objects by keeping the fingers horizontally. We assume that the fingers should translate and/or rotate in a horizontal plane (so-called *grasping plane*), which is a virtual copy of the domestic plane with a certain vertical offset. The specific point P_3 belongs to this grasping plane and is used as the starting position for incrementally reaching a grasping situation at the target object.

The point P_3 should be defined advantageous with regard to the *problem of obstacle avoiding*, *e.g.* it is convenient to construct the point as follows. We determine a straight line between a center point of the robot arm (*e.g.* origin of coordinate system) and the position of the target object, vertically project this line onto the grasping plane, and take the intersecting boundary point of the grasping plane, which is nearest to the robot arm, as the point P_3 (see top view from robot arm and domestic area in Figure 4.12). The target object can be grasped only if there is a enough free space along the route of approaching the object. Generally, the space between the robot hand and the robot center is occupied to a certain extent by the robot arm and its 3D volume is non-rigid. In consequence of this, the system must determine a route to the target object such that the requested 3D space for the robot arm is not occupied by obstacle objects (see Figure 4.13).

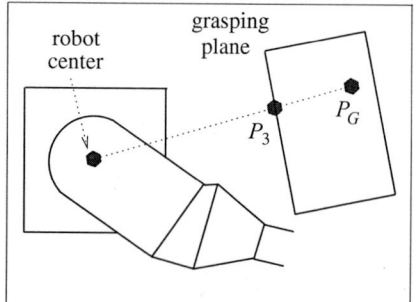

Fig. 4.12. Constructing point P_3 based on center position of robot arm and position P_G of target object.

It may happen that objects are located quite densely and no collision-free route to the target object is left. In this case, the relevant obstacle objects must be *determined* and *carried* to the parking area. In our application, the two sub-tasks are sequentialized completely, i.e first determining all relevant obstacles and second carrying them away (the latter is treated later on). It is favourable to determine potential obstacle objects by moving the hand-camera CA_2 over the domestic area with the optical axis directed vertically. Based on the collection of single images one obtains a global impression from the arrangement of all objects which is necessary for planning the collision-free route automatically. More concretely, we only need to know those objects which are located between the target object and the robot arm, and the other objects are not relevant, because they are not on the path of approaching the target. Based on the global impression the system should determine those

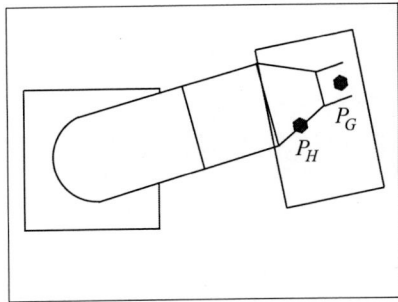 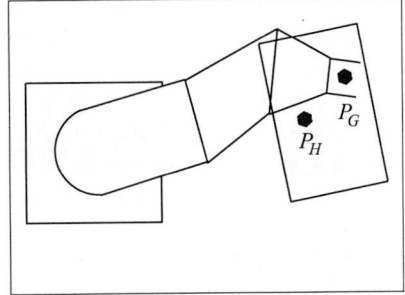

Fig. 4.13. (Left) Grasping a target object at position P_G and collision with obstacle object at position P_H; (Right) Grasping without obstacle collision.

route to the target object along which a minimum number of obstacles is located, *i.e. carrying away only a minimum number of obstacles.*

For planning the collision-free route we take the non-rigid 3D volume of the robot arm into account, which changes continually during the movement of the robot hand. In addition to this, we must generate an approximation for the three-dimensional shape of the set of potentially relevant objects. The non-rigid volume of the robot arm and the approximated object volumes must not overlap. Consequently, during the movement of the hand-camera CA_2 over the domestic area we must take images according to a strategy such that it is possible to reconstruct 3D information from the relevant objects. The *structure-from-motion-stereo* paradigm can be applied for 3D reconstruction, *i.e.* the hand-camera CA_2 moves in small steps and takes images, 2D positions of certain image features are extracted, correspondences between the features of consecutive images are determined, and finally 3D positions are computed from 2D correspondences. Essentially, for simplifying the correspondence problem we must *take images with small displacements*, which is quite different from the strategy for localizing the target object (see Subsection 4.3.2).

A specific version of the structure-from-motion-stereo paradigm is implemented by Blase [22].[13] The 3D shape of an obstacle object is approximated on the basis of a collection of points which originate from the surface of the object. In the images these points are determined as gray value corners and for their detection the *SUSAN operator* is applied. Correspondences of gray value corners between consecutive images are obtained by normalized cross correlation of small patterns which are centered at the extracted features.

For example, Figure 4.14 shows corresponding gray value corners between consecutive images. The function for *3D reconstruction from two corresponding 2D positions* is approximated by a mixture of radial basis function networks which must be trained with the use of a calibration pattern. In the

---

[13] Diploma thesis (in german) supervised by the author of this book.

application phase the *camera moves along in front of a collection of objects*, and during this process some of the objects appear in the field of view and others may disappear (see Figure 4.15). Continually, 3D surface points are reconstructed from pairs of 2D positions of corresponding gray value corners. The collection of 3D points must be cleared up from outliers and clustered according to coherence (see left and right image in Figure 4.16). The convex hull of each cluster is used as an approximation of the 3D volume of the scene object. The approach is only successful if there is a dense inscription or texture on the surface of the objects (*e.g.* inscriptions on bottles). Otherwise, a more sophisticated approach is necessary, *e.g.* extracting object boundaries from images by using techniques presented in Chapter 2.

Fig. 4.14. Extracting gray value corners in two consecutive images; detecting correspondences.

Fig. 4.15. Four consecutive images and extracted gray value corners (black dots).

3D boundary approximations of objects are the basis for *planning collision-free routes* over the domestic area. In a previous sub-task the target object has been localized in the image and discriminated from all other objects. Consequently, we can distinguish the 3D boundary of the target object from the 3D boundaries of the other objects. On the other hand, we need to know the *volume of the robot arm* which is different for each position of the robot hand. This volume can be determined by computing the inverse kinematics and, based on this, taking the fixed state vector of the robot arm into account, *e.g.* length and diameter of links. We decided that the approach-

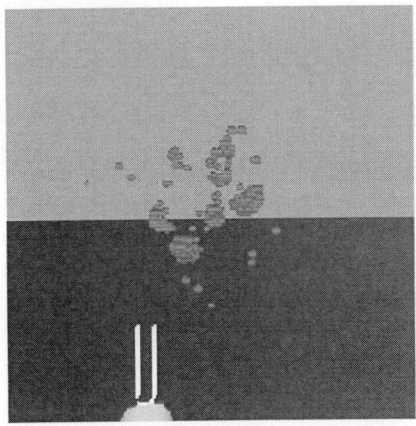

Fig. 4.16. (Left) Collection of 3D points originating from the bottles in Figure 4.15; (Right) Clusters of 3D points cleared up from outliers.

ing of the target object must be carried out on a horizontal plane to which point P_3 belongs to, *i.e.* this point is the starting position of the movement (see Figure 4.9). A possible planning strategy for obtaining a collision-free route to the target object works with the methodology of dynamical systems. Concretely, the following six steps are involved.

First, the three-dimensional boundary shapes of all relevant objects are projected vertically on the grasping plane, which results in silhouettes of the objects viewed from the top. Second, we specify an attractor on the grasping plane at the center position of the silhouete of the target object, *i.e.* we assume this is the place from where to grasp the target object. Third, we spread out a set of repellors equidistantly over the silhouette boundaries of each other object, respectively. This multitude of repellors for each potential obstacle object is useful for keeping the robot hand off from any part of the object surface. Fourth, the attractor and repellor vector fields are summed up. From the resulting field we can extract movement vectors leading towards the target object, if there is a possible route at all. However, so far we did not take into account the volume of the robot arm. Fifth, for this purpose we apply an exploration approach, which will take place in *virtual reality* including virtual movements of the robot hand. Based on the previously determined movement vectors the robot hand is moving virtually along a suggested vector (beginning from point P_3), and for the new position it is tested whether a virtual collision with an object has occured. If this was the case, then a repellor is specified for this hand position, and the hand is moved back virtually and another hypothetical movement vector can be tried again. Sixth, if there is no collision-free route towards the target object we determine the object which is located on or near the straight line between position P_3 and the position of the target object and is most closest to P_3. This object is considered as obstacle which must be carried to the parking area.

The original sub-task of approaching the target object must be defered for a moment. The new sub-task of approaching the obstacle object is reflected in the *vector field representation* as follows. We take a copy from the original vector field, erase the effector at the position of the target object, erase all repellors at the relevant obstacle object (determined as discussed above), and specify an attractor at the position of this relevant obstacle object. Based on the updated superposition of attractor and repellor vector fields the robot hand can approach the obstacle object for grasping it finally.

Task-Specific Modules for Determining/Reconstructing Obstacles

As a result of the experimental designing phase we define two task-specific modules which must be applied in the application phase sequentially.

The first task-specific module MT_2 is responsible for the meander-type movement over a part of the domestic area and taking a series of images (generic module MI_3), reconstructing the boundary of target object and obstacle objects, determining point P_3, moving the robot hand to position P_3 and rotating robot fingers parallel to grasping plane (generic module MI_2).

$$MT_2 \quad := \quad (\ MI_3 \quad \& \quad MI_2 \ ) \tag{4.9}$$

The meander-type movement is restricted to the sub-area between robot center and target object, and images are taken in small incremental steps. The pre-processing function must determine the meander structure on the viewing plane over a part of the domestic area, and define many intermediate stops (as discussed above). Post-processing functions are responsible for the extraction of object boundaries, detection of correspondences, 3D reconstruction, projection on grasping plane, and determining point P_3 on the grasping plane. The output of the module comprises the silhouette contours of target object and the other objects, and the starting point P_3 on the grasping plane.

The second task-specific module MT_3 is responsible for determining the obstacle object (located nearest to point P_3) which prevents the robot hand from approaching the target object. This module is different from the previous task-specific modules in that no real but only virtual movements of the robot hand will take place, *i.e.* it is a task-specific *planning* module. The overall vector field must be updated by doing virtual exploratory movements for avoiding collisions with robot body. A pre-processing function must construct an overall vector field from silhouette contours of target object and obstacle objects. Post-processing functions determine the relevant obstacle object, modify the vector field such that the obstacle object can be approached. The outcome is a vector field for approaching the obstacle object.

4.3.4 Approaching and Grasping Obstacle Objects

In the application phase the *third goal* is to approach and grasp an obstacle object.[14]

Designing Aspects for Approaching/Grasping Obstacles

The obstacle object has been determined such that during the approaching process no collision will occur with other objects. Starting at position P_3 the obstacle object can be approached by following the force vectors until the equilibrium (attractor center) is reached. In simple cases, an instructional module of type MI_2 is applicable which executes an assembled instruction without continual visual feedback. The usefulness of this strategy is based on the assumption that the position of the obstacle object can be reconstructed exactly and that the object can be grasped arbitrary. This assumption does not hold in any case, and therefore we present a more sophisticated strategy which combines deliberate plan execution with visual feedback control.

The first part of the strategy is equal to the instructional movement as mentioned just before, *i.e.* the robot hand will approach the obstacle object by following the *vectors of the deliberate vector field*. However, the plan must be executed merely as long as the obstacle object is not contained in the field of view of hand-camera CA_3 or is just partly visible. The hand-camera CA_3 is supposed to be used for a fine-tuned control of the grasping process, and for this purpose both the grasping fingers and the grasping object must be located in the field of view. As soon as the object is visible completely, the plan is interrupted and a process of *visual feedback control* continues with the sub-task of grasping. The robot hand must be carefully servoed to an optimal grasping situation, *i.e.* a high accuracy of assembling is desired. For this purpose, the hand-camera CA_3 must be equipped with an objective such that both robot fingers and at a certain distance the grasping object can be depicted at a reasonable resolution.

It is assumed that the silhouette of the object is elongated (instead of round) and the length of the smaller part is less than the distance between the two grasping fingers, *i.e.* grasping is possible at all. For a detailed explanation of the servoing strategy we introduce for the robot hand a *virtual hand axis* and a *virtual gripper point*. The virtual hand axis is the middle straight line between the two elongated fingers. The virtual gripper point is obtained by first computing the end straight line, which connects the two finger tips, and then intersecting this line with the virtual hand axis. Furthermore, we also introduce a *virtual object axis* of the grasping object which is defined as the first principal component axis of the object silhouette.

In Subsection 3.4.1 we discussed about strategies for an efficient recognition of objects or situations, *e.g.* reducing the complexity of the manifold of appearance patterns. For keeping the manifold of grasping situations

---

[14] Sophisticated work on automated grasp planning has been done by Röhrdanz *et al.* [142].

tractable we found it favourable to *sequentialize the grasping process* into four phases (see Figure 4.17). First, the robot hand must reach a situation such that the virtual hand axis is running through the center position of the object silhouette. Second, the robot hand should *move perpendicular to the virtual hand* axis until the virtual gripper point is located on the virtual object axis. Third, the robot hand must *rotate around the virtual gripper point* for making the virtual hand axis and the virtual object axis collinear. Fourth, the robot hand should *move along the virtual hand axis* until an optimal grasping situation is reached. All four phases must be executed as servoing procedures, including continual visual feedback, to take care for inaccuracies or unexpected events. Figure 4.18 shows the intermediate steps of the grasping process in real application. In the following we present examples for the type of measurements in the images and for the control functions on which the behavioral modules are based.

Four Phases of a Robotic Grasping Process

The aspects of the *first phase* of the grasping process are included more or less in the successive phases and therefore is not discussed specifically.

For the *second phase* we may work with the virtual gripper point and the virtual object axis explicitly. The virtual gripper point can be extracted by a combination of gray value and geometric features as follows. By normalized cross correlation the gripper tip is located roughly (see Figure 4.19). In order to verify the place of maximum correlation and localize the position of the virtual gripper point exactly we additional extract geometric features of the fingers.

Hough transformation can be used for extracting the elongated straight lines of the finger boundaries. Under the viewing perspective of the hand-camera CA_3 the two top faces of the fingers are visible clearly and appear brightly, and the two silhouettes of the top faces mainly consist of two elongated lines, respectively. Taking the polar form for representing lines, the Hough image is defined such that the horizontal axis is for the radial distance and the vertical axis is for the orientation of a line. According to this, the two pairs of elongated boundary lines of the parallel jaw gripper occur in the Hough image as four peaks which are nearly horizontal due to similar line orientations (see Subsection 2.3.1). According to these specific pattern of four peaks the elongated finger lines are extracted and from those also the virtual hand axis. The end straight line which connects the two fingers tips virtually, can be extracted by using Hough transformation in combination with the SUSAN corner detector (see Subsection 2.2.3). The virtual gripper point is determined by intersecting the relevant lines (see Figure 4.20). On the other hand, we will extract the virtual axis of the grasping object as the first principal component axis of the object silhouette.

Based on all these features, we can define the measurement vector $Q(t)$ for the servoing procedure as the euclidean distance between the virtual gripper

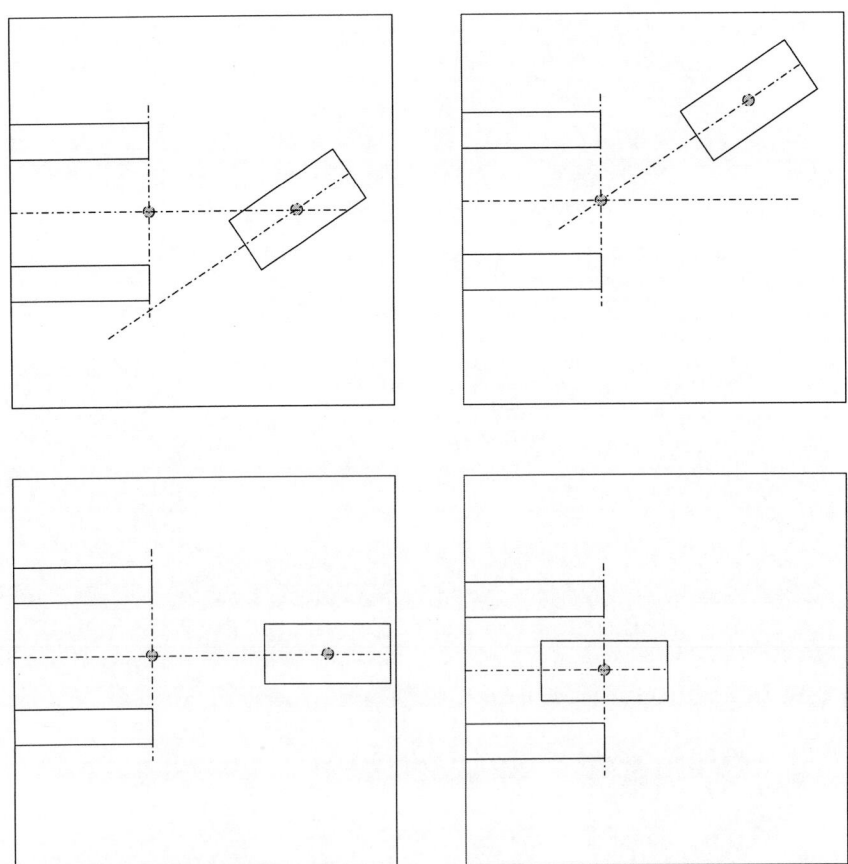

Fig. 4.17. Grasping process organized in three phases, *i.e.* perpendicular, rotational and collinear phase of movement.

point and the virtual object axis normal to the virtual hand axis. The desired measurement vector Q^* is simply the value 0. The control vector for moving the robot hand is biased, *i.e.* leaving only one degree of freedom for moving the robot hand on the grasping plane perpendicular to the virtual hand axis. Especially, both the measurement vectors and the control vectors are scalar. A constant increment value s is prefered for easy tracking the movement, and a reasonable value is obtained in the experimental designing phase. That is, a Jacobian must be determined which describes the relationship between displacements of the robot hand and the resulting displacements in the image of hand-camera CA_3 (see later on Subsection 4.4.2). In this special case the Jacobian is a trivial matrix containing just the constant value s. Then, the control function is defined as follows.

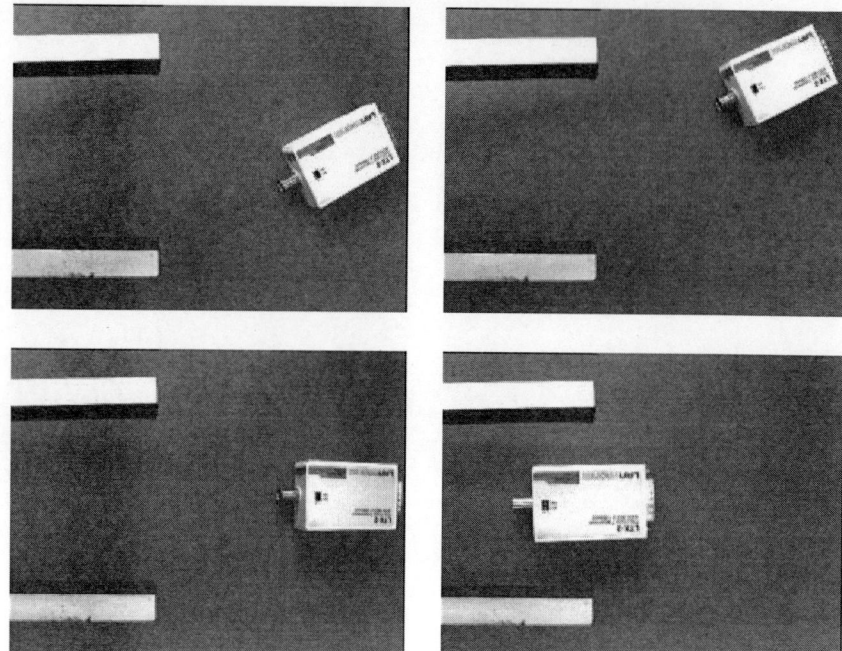

Fig. 4.18. Assembling the gripper to an object.

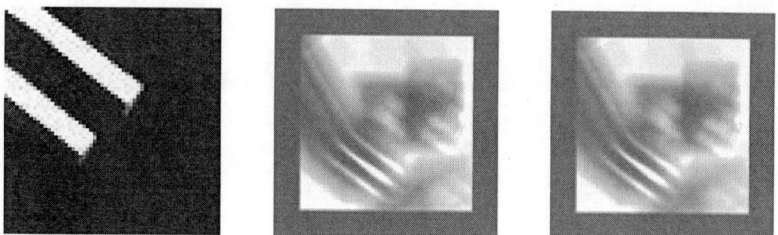

Fig. 4.19. Gripper, gripper tip region, correlation image.

$$C(t) \quad := \quad \begin{cases} s \cdot signum(Q^* - Q(t)) & : \quad |Q^* - Q(t)| > \eta_1 \\ 0 & : \quad else \end{cases} \qquad (4.10)$$

Parameter η_1 specifies the acceptable deviation from desired value 0.

In the *third phase* of the grasping process the robot hand should rotate around the virtual gripper point for reaching collinearity between the virtual hand axis and the virtual object axis. Just one degree of freedom of the robot hand must be controlled for rotating in the grasping plane, and the remaining state variables of the robot hand are constant during this sub-task. The type of measurements in the image can be equal to the previous sub-task of perpendicular hand movement, *i.e.* extracting virtual hand axis and

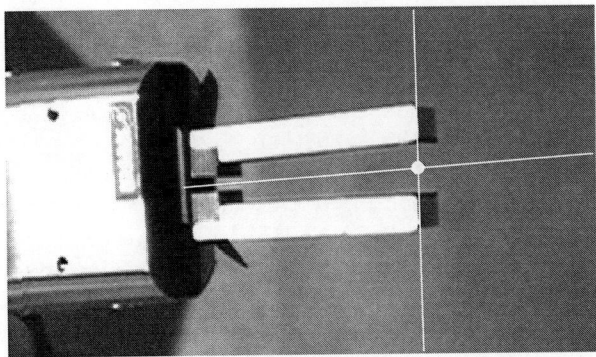

Fig. 4.20. Construction of finger boundary lines, virtual hand axis, end straight line, and virtual finger point.

virtual object axis. Based on this, we compute the angle between both axes for specification of a scalar measurement vector $Q(t)$, which should reach the desired value 0 finally. The control function is equal to the one presented in equation (4.10). Instead of extracting geometric features explicitly, we briefly mention two other types of measurements in the image which characterize the orientation deviation more implicitly.

The *log-polar transformation* can be applied for tracking the obstacle object in the LPT image by techniques of cross correlation (see explanations in Subsection 3.4.2). The camera executes a simple rotational movement around the virtual gripper point, which implies the impression that the object is rotating around this point (see in Figure 4.17 the images on top right and bottom left). If we compute log-polar transformation continually with the center of the foveal component defined as the virtual gripper point, then the LPT pattern is translating along the θ axis of the polar coordinate system (see exemplary the Figures 3.21 and 3.22). Cross correlation is applicable for localizing the LPT pattern of the grasping object. Due to the incremental movements the relevant search area can be constrained which is useful for reasons of efficiency. The virtual hand axis and virtual object axis are approximately collinear, if and only if the LPT pattern is located on the vertical axis of the LPT image defined by $\theta = 0$. Therefore, alternatively to the previous type the scalar measurement vector $Q(t)$ can be defined also by the current value of θ, which is supposed to represent the center of the LPT pattern of the obstacle object along the horizontal axis of the LPT image. That is, this value of θ serves as a measurement of the orientation deviation.

Alternatively to the approaches of extracting virtual axes or the log-polar transformation, we can determine the orientation deviation based on histograms of edge orientations. The approach is easy if the background is homogeneous and the field of view only contains the fingers and the grasping object. In this case the original image can be transformed into a binary image representing gray value edges. We perserve the edges orientations and construct a histogram thereof. The left and right diagram in Figure 4.21 show these histograms prior and after the cycles of rotational gripper movement (for second and third image in Figure 4.18). The position of the first peak in Figure 4.21 (left), which is close to 65°, specifies the principal orientation ϕ^{ob} of the object and the second larger peak (*i.e.* close to 90°) the gripper orientation ϕ^{gr} in the image. During the servoing cycle the gripper orientation changes but due to the fastened camera a change of the object orientation appears. Accordingly, the first histogram peak must move to the right until it unifies into the second peak (Figure 4.21, right). The current measurement vector $Q(t)$ is defined by ϕ^{ob}, and the desired measurement vector Q^* by ϕ^{gr}.

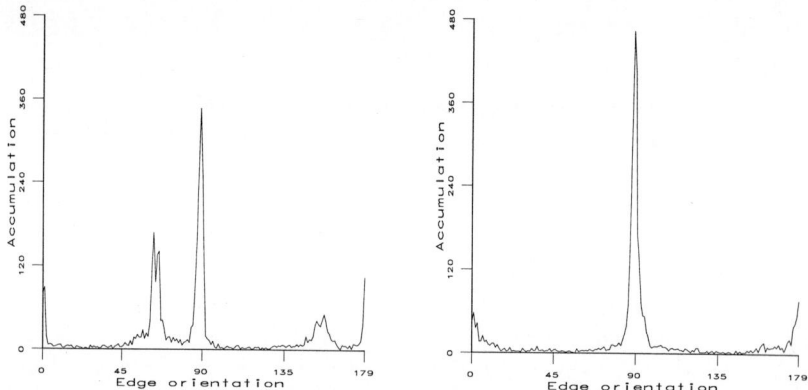

Fig. 4.21. Edge orientation histograms; (Left) Prior to finger rotation; (Right) After finger rotation.

In the *fourth phase* of the grasping process the robot hand should move collinear with the virtual hand axis in order to reach an optimal grasping situation. For defining grasping situations we can take the virtual gripper point and the object center point into account, *e.g.* computing the euclidean distance between both. If this distance falls below a certain threshold, then the desired grasping situation is reached, else the gripper translates in small increments.

An alternative approach for evaluating the grasping stability has been presented in Subsection 3.3.4 which avoids the use of geometric features. A GBF network learns to evaluate the stability of grasping situations on the basis of training examples. Those example situations are represented as patches of filter responses in which a band pass filter is tuned to respond specifically on certain relationships between grasping fingers and object. The filter responses represent implicitly a measurement of distance of the gripper from the most stable position. For example, if the gripper moves step by step to the most stable grasping pose and then moves off, and sample data are memorized thereof, then the network may learn a parabolic curve with the maximum at the most stable situation. A precondition for applying the approach is that gripper and object must be in a small neighborhood so that the filter can catch the relation.

Instead of computing for the vector of filter responses a value of grasping stability it is possible to associate an appropriate increment vector for moving the gripper. In this case, the control function is implemented as a neural network which is supposed to be applied to a filter response vector. We do not treat this strategy in more detail.

Task-Specific Modules for Grasping Obstacle Objects

As a result of the experimental designing phase we define a task-specific module MT_4 which is based on the execution of a generic module of type MB_3, followed by the sequential execution of four generic modules each of type MB_1, followed by an elementary instruction of type MI_2.

$$MT_4 \quad := \quad (\ MB_3 \ \& \ MB_1 \ \& \ MB_1 \ \& $$
$$MB_1 \ \& \ MB_1 \ \& \ MI_2 \ ) \qquad\qquad (4.11)$$

The module of type MB_3 is responsible for approaching the robot-hand to the obstacle object by following the vectors of the deliberate vector field. The plan is interrupted as soon as the obstacle object is completely visible in the field of view of the hand-camera CA_3. The four generic modules of type MB_1 implement elementary behaviors, *i.e.* the first one is responsible for arranging the robot hand such that the virtual hand axis is running through the center position of the object silhouette, *i.e.* the second one is responsible for the perpendicular translation of the robot hand relative to the virtual hand axis, the third one is responsible for the relevant rotation, and the fourth one for collinear translation for reaching the optimal grasping situation (see above). The instructional module of type MI_2 implements the closing of the fingers for grasping.

4.3.5 Clearing Away Obstacle Objects on a Parking Area

In the application phase the *fourth goal* is to clear away obstacle objects on a parking area.

Designing Aspects for Clearing Away Obstacle Objects

The robot hand should *lift up* the obstacle object a pre-specified distance and *move it backward* step by step to a virtual point above starting position P_3, *i.e.* position P_3 modified in the Z-coordinate by the pre-specified distance. Then, the object must be *carried* from the domestic area to the parking area along a *pre-specified trajectory*. The goal position of this trajectory is a virtual point P_4 located a certain distance above the parking area. Finally, the object must be carefully *servoed to a certain place* on the parking area in order to *put down the object*. For solving these three sub-tasks we must take other cameras into account for grounding processes of visual feedback control. The hand-cameras CA_2 and CA_3 are no more useful because the fields of view are occupied to a large extent by the grasped object, and furthermore this object occludes the space behind it.

The *head-cameras* CA_4 and CA_5 are intended for the visual feedback control of *approaching the object to a free place* on the parking area. We assume that a relevant free place has already been found (*i.e.* do not treat this sub-task), and that the virtual point P_4 has been determined as an appropriate starting position for approaching this place. The head-cameras must be aligned appropriately using the pan, tilt, and vergence degrees of freedom, *e.g.* the optical axes should be directed to the center of the parking area. The zooming/focusing degree of freedom should be tuned such that the area and the position from where to start the approaching process are completely contained in the field of view, at the maximal possible resolution, and under the optimal focus. In a later sub-task the head-cameras CA_4 and CA_5 will also be used to control the process of approaching an object to a place on the inspection area. We assume that the double usage of the head cameras is possible without changing the position or orientation of the robot vehicle (containing the robot head). Instead, for solving the inspection task (later on) only the degrees of freedom of robot head and cameras are supposed to be changed appropriately, *i.e.* keeping similar viewing constraints for the inspection area which have been mentioned for the parking area. The domestic area and inspection area are fixed and therefore the robot vehicle can be steered into the relevant position prior to the application phase. Relative to this fix vehicle position the two relevant state vectors of the head-camera system can be determined for optimally observing the parking area and the inspection area, respectively.[15]

The head-cameras must take images continually for the visual feedback control of putting down an object on a goal place of the parking area. In each of the stereo images both the object and the goal place are visible, which is a precondition for determining a certain kind of distance. Based on the distance measurements in the two images, a control vector is computed for carrying the object nearer to the goal place. This principle can also be applied for treating the peg-in-hole problem, *e.g.* in a system implemented by Schmidt we used cyclinders and cuboids [150].[16] The critical issue is to extract the relevant features from the stereo images.

---

[15] Task-specific modules are needed for supporting the system designer in the sub-task of vehicle and camera alignment. However, in this work we don't care about that.

[16] Diploma thesis (in german) supervised by the author of this book.

For example, let us assume a *cylindrical object and a circular goal place* as shown in the top left image of Figure 4.22. The binary image which has been computed by thresholding the gradient magnitudes is depicted on top right. In the next step, a specific type of Hough transformation is applied for approximating and extracting half ellipses. This specific shape is supposed to occur at the goal place and at the top and bottom faces of the object. Instead of full ellipses we prefer half ellipses, concretely the lower part of full ellipses, because due to the specific camera arrangement that feature is visible throughout the complete process. From the bottom face of the object only the specific half ellipse is visible. The process of approaching the object to the goal place is organized such that the lower part of the goal ellipse remains visible, but the upper part may become occluded more and more by the object. The extraction of the lower half of ellipses is shown in the bottom left image in Figure 4.22. The distance measurement between object and goal place just takes the half ellipse of the goal place and that from the bottom face of the object into account. For computing a kind of distance between the two relevant half ellipses we extract from each a specific point and based on this we can take any metric between 2D positions as distance measurement. The bottom right image in Figure 4.22 shows these two points, indicated by crosses, on the object and the goal place.

The critical aspect of extracting points from a stereo pair of images is that *reasonable correspondences* must exist. A point of the first image is in correspondence with a point of the second image, if both originate from the same 3D point. In our application, the half ellipses extracted from the stereo images are the basis for determining corresponding points. However, this is by no means a trivial task, because the middle point of the contour of the half ellipse is not appropriate. The left picture of Figure 4.23 can be used for explanation. A virtual scene consists of a circle which is contained in a square. Each of the two cameras produces a specific image, in which an ellipse is contained in a quadrangle. The two dotted curves near the circle indicate that different parts of the circle are depicted as lower part of the ellipse in each image. In consequence of this, the middle points p_1 and p_2 on the lower part of the two ellipses originate from different points P_1 and P_2 in the scene, *i.e.* points p_1 and p_2 do not correspond. Instead, the right picture of Figure 4.23 illustrates an approach for determining corresponding points. We make use of a specific geometric relation which is invariant under geometric projection.[17]

We translate, virtually, the bottom line of the square to the circle which results in the tangent point P. This procedure is repeated in the two images, *e.g.* translating the bottom line of the quadrangle parallel towards the ellipse

---

[17] In Subsection 1.4.1 we discussed about compatibilities of regularities under geometric projection. They have been proven advantageous in Chapter 2 for extracting object boundaries. In this subsection we present another example for a compatibility of regularities under geometric projection. It will be useful for extracting relevant image features from which to determine correspondences.

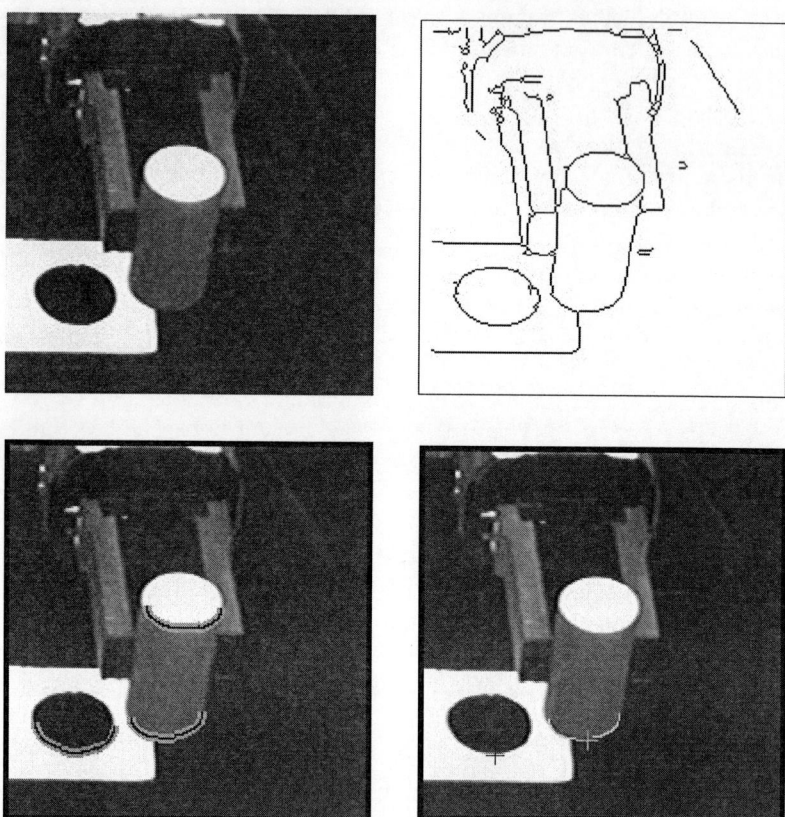

Fig. 4.22. (Top left) Cylinder object and a circular goal place; (Top right) Binary image of thresholded gradient magnitudes; (Bottom left) Extracted half ellipses; (Bottom right) Specific point on half ellipses of object and goal place.

to reach the tangent points p_1 and p_2. Due to different perspectives the two bottom lines have different orientations and therefore the resulting tangent points are different from those extracted previously (compare left and right picture of Figure 4.23). It is observed easily that the new tangent points p_1 and p_2 correspond, *i.e.* originate from the single scene point P. In our application, we can make use of this *compatibility*. For this purpose one must be careful in the experimental designing phase to specify an appropriate state vector for the head-camera system.

Especially, the parking area (including the boundary lines) must be completely contained in the field of view of both head-cameras. In the application phase a certain part of the boundary lines of the parking area is extracted from the two images (the stereo correspondence between those specific image lines can be verified easily). For each image the orientation of the respective boundary line can be used for determining relevant tangent points at the rel-

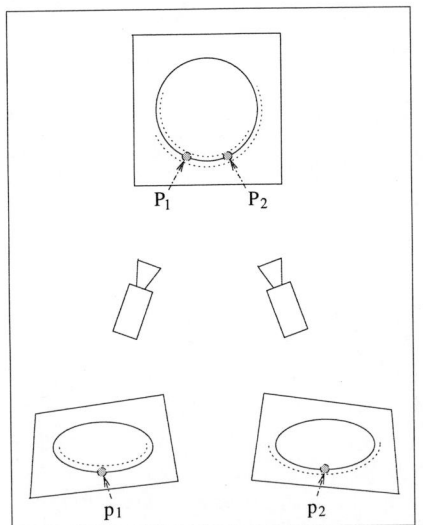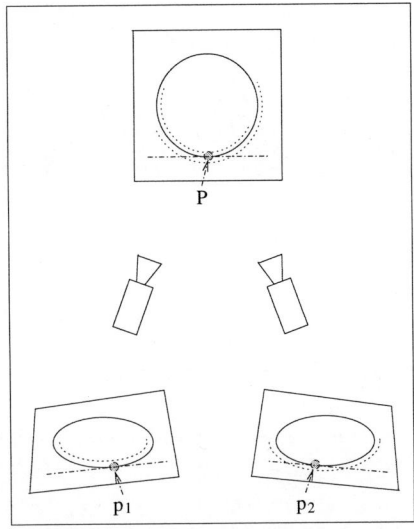

Fig. 4.23. (Left) Extracted image points p_1 and p_2 originate from different scene points P_1 and P_2, (Right) Extracted image points are corresponding, *i.e.* originate from one scene point P.

evant ellipse, *i.e.* virtually move the lines to the ellipses and keep orientation. Tangent points must be extracted at the half ellipse of the goal place and at the half ellipse of the bottom face of the object. These points have already been shown in the bottom right image of Figure 4.22.

For defining the control vector we need to describe the relationship between displacements of the robot hand and the resulting displacements in the two stereo images taken by the head-cameras. For this purpose we introduce two Jacobians $J_1^f(P)$ and $J_2^f(P)$ which depend on the current position P of the virtual gripper point. If we multiply the Jacobian $J_1^f(P)$ (respectively Jacobian $J_2^f(P)$) with a displacement vector of the hand position, then the product will reveal the displacement vector in the left image (respectively in the right image). The two Jacobians are simply joined together which results in a (4×3) matrix depending on P.

$$J^f(P) := \left(\begin{array}{c} J_1^f(P) \\ J_2^f(P) \end{array} \right) \tag{4.12}$$

In order to transform a desired change from stereo image coordinates into manipulator coordinates the pseudo inverse $J^\dagger(P)$ is computed.

$$\left(J^f\right)^\dagger (P) := \left(\left(J^f\right)^T (P) \cdot \left(J^f\right)(P) \right)^{-1} \cdot \left(J^f\right)^T (P) \tag{4.13}$$

The current position $P(t)$ of the virtual gripper point defines the variable state vector $S^v(t)$. The desired measurement vector Q^* is a 4D vector comprising the 2D positions of a certain point of the goal place in the stereo images. The current measurement vector $Q(t)$ represents the stereo 2D positions of a relevant point on the object (see above).

$$Q^* := \begin{pmatrix} p_1^* \\ p_2^* \end{pmatrix} ; \quad Q(t) := \begin{pmatrix} p_1(t) \\ p_2(t) \end{pmatrix} \tag{4.14}$$

With these definitions we can apply the following control function.

$$C(t) := \begin{cases} s \cdot \left(J^f\right)^\dagger \left(S^v(t)\right) \cdot \left(Q^* - Q(t)\right) & : & |Q^* - Q(t)| > \eta_1 \\ 0 & : & else \end{cases} \tag{4.15}$$

with the servoing factor s to control the size of the steps of approaching the goal place. The hand position is changed by a non-null vector $C(t)$ if desired and current positions in the image deviate more than a threshold η_1. Actually, equation (4.15) defines a proportional control law (P-controller), meaning that the change is proportional to the deviation between the desired and the current position.[18]

Task-Specific Modules for Clearing Away Obstacle Objects

As a result of the experimental designing phase we define a task-specific module MT_5 which is based on the simultaneous execution of two generic modules of type MI_2, followed by a generic module of type MB_1, followed once again by a simultaneous execution of two basic modules of type MI_2.

$$MT_5 \quad := \quad ((\ MI_2 \quad | \quad MI_2 \ ) \quad \& \quad MB_1 \quad \& $$
$$(\ MI_2 \quad | \quad MI_2 \ )) \tag{4.16}$$

The first module of type MI_2 is responsible for lifting up the obstacle object a pre-specified distance and move it backward step by step to a virtual point above starting position P_3, and from there carry the object to a virtual point P_4 located a certain distance above the parking area. Simultaneously, another module of type MI_2 can be executed for changing the degrees of freedom of the head system such that the parking area is located in the field of view. Next, the module of type MB_1 can be started, which is responsible for the elementary behavior of putting down the object at the goal place by visual feedback control. Finally, a module of type MI_2 is responsible for opening the robot fingers, lifting up a certain extent, and moving the robot hand backward to position P_2 of the domestic area. Simultaneously, another module of type MI_2 can be executed for changing the degrees of freedom of the head system such that the original area is located in the field of view.

---

[18] Alternatively, the P-controller can be combined with an integral and a derivative control law to construct a PID-controller. However, the P-controller is good enough for this simple control task.

4.3.6 Inspection and/or Manipulation of a Target Object

The *fifth* goal in our series of sub-tasks is the inspection and/or manipulation of a target object. For obtaining a collision-free route (on the domestic area) towards the target object one must repeat the relevant task-specific modules several times. These are task-specific module MT_3 for determining an obstacle object, the module MT_4 for approaching and grasping an obstacle object, and the module MT_5 for clearing away the obstacle object on the parking area.

$$MT_6 \quad := \quad (MT_3 \ \& \ MT_4 \ \& \ MT_5)^* \qquad\qquad (4.17)$$

The * symbol denotes a repetition of the succession of the three sub-tasks. In the case that no further object prevents the robot hand from approaching the target object, one can apply module MT_4 for approaching and grasping the target object, and finally apply a module which is similar to MT_5 for carrying the target object to the inspection area. We do not treat this anew, because only minor changes are required compared to the previous specifications, *e.g.* carrying the target object to the virtual position P_5 and then putting it down to the center of the inspection area by visual feedback control. According to this, we assume that the target object is already located on the inspection area.

There is a wide spectrum of *strategies for inspecting objects*, but we concentrate only on *two approaches*. First, we discuss criteria for evaluating viewing conditions in order to obtain one *optimal image from an object*. Second, we present an approach for continual handling of an effector, which may carry a camera, for the *visual inspection of large objects*.

Designing Aspects for Reaching Optimal Viewing Conditions

For object inspection more detailed information about the target object must be acquired. We would like to take an image such that the object appears with a reasonable size or resolution, at a certain level of sharpness, and under a specific orientation. It is assumed that the target object is located at the center of the inspection area and the optical axes of the two head-cameras are aligned to this point. However, for the specific sub-task only one head-camera is used. For changing the size, resolution, or sharpness of the object appearance we must fine-tune the focal length or the lens position of the head-camera. In addition to this, appropriate object orientation is reached by controlling the angle of the rotary table. Figure 4.24 shows an object taken under large and small focal length (left and middle image), and under degenerate orientation (right image).

The change of the depicted object size can be evaluated by *image subtraction, active contour construction, optical flow computation, etc.* For example, an active contour approach [176] is simply started by putting an initial contour at the image center and then expanding it step by step until the background image area of the object is reached which is assumed to be homogeneous. Based on this representation it is easy evaluated whether the object

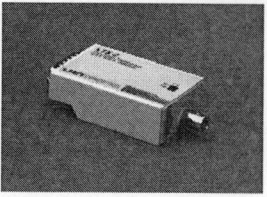

Fig. 4.24. Transceiver box, taken under large and small focal length, and under degenerate orientation.

silhouette is of a desired size or locally touches the image border and thus meets an optimality criterion concerning depicted object size. In addition to this, the sharpness of the depicted object should surpass a pre-specified level. A measure of sharpness is obtained by computing the magnitudes of gray value gradients and taking the mean of a small percentage of maximum responses. High values of gradient magnitudes originate from the boundary or inscription of an object, and low values originate from homogenous areas. However, sharpness can best be measured at the boundary or inscription. Therefore, the critical parameter which determines a percentage of maximum responses, describes the assumed proportion of boundary or inscription of the object appearance relative to the whole object area. In the designing phase, this parameter must be specified, which typically is no more than 10 percent. The measurements of sharpness should be taken within the silhouette of the object including the boundary.[19]

The change of object resolution in the image can be evaluated by frequency analysis, Hough transformation, steerable filters, *etc.* For example, by using Hough transformation we extract boundary lines and evaluate distances between approximate parallel lines. A measure of resolution is based on the *pattern of peaks* within a horizontal stripe in the Hough image. Figure 4.25 shows for the images in Figure 4.24 the Hough image, respectively. For the case of low (high) resolution the horizontal distances between the peaks are small (large). Having the object depicted at the image center the straight boundary lines of a polyhedral object can be approximated as straight image lines due to minimal perspective distortions. Maybe, in the previous phase of localizing the target object on the domestic area, the reliability of recognition was doubtful (see Subsection 4.3.2). Now, having this object located on the inspection area, one can identify it more reliablly by taking images under a general object orientation. For example, three visible faces of the transceiver box in Figure 4.24 (left and middle) are more useful than the degenerate object view in Figure 4.24 (right) which shows only two faces. Taking the peak pattern of the Hough transformation into account we can differentiate between general and degenerate views (see Figure 4.25, middle and right).

---

[19] Experiments on measurements of sharpness are presented in Subsection 4.4.4 (later on).

According to this, the object can be rotated appropriately while preserving its position on the rotary table.

Fig. 4.25. Hough transformation of binarized images in Figure 4.24.

Task-Specific Modules for Reaching Optimal Viewing Conditions

We discussed only superficially the designing-related aspects for reaching optimal viewing conditions, and in a similar way we will discuss about relevant task-specific modules.[20] The generic module of type MB_1 could be useful for changing the focal length incrementally with the goal of reaching a desired size for the object appearance in the image. Maybe, the control of the focal length must be accompanied with a control of the lens position for fine-tuning the sharpness. In this case a generic module of type MB_2 would be appropriate for treating two goals in combination. However, in general it is difficult to specify goals explicitly which may represent optimal viewing conditions, *e.g.* it is difficult to describe the optimal object orientation. According to this, basic behavioral modules are useful which do not rely on explicit goals (*e.g.* desired measurements) and instead implement exploratory strategies, *e.g.* a generic module of type MB_3. Depending on specific applications different exploratory behaviors are requested, however we do not treat this in more detail.

Designing Aspects for Continual Handling of an Effector

The effector may carry a camera for the visual inspection of large objects. Alternatively, the effector may also be configured as a tool which could be used for incremental object manipulation. In the following we treat both alternative applications in common. Generally, it is required to move the effector, *e.g.* the robot hand in our robot system, along a certain trajectory and furthermore *keep a certain orientation* relative to the object. For example, we assume that a gripper finger must be servoed at a certain distance over an object surface and must be kept normal to the surface.

In the following, we take the application scenery of dismantling computer monitors. A plausible strategy is to detach the front part of a monitor case using a laser beam cutter. The trajectory of the cutter is approximately a

---

[20] The sub-task of reaching optimal viewing conditions requires a more principled treatment which is beyond the scope of this work.

rectangle (not exactly a rectangle) which surrounds the front part at a constant distance, and during this course the beam orientation should be kept orthogonal to the relevant surface part. Figure 4.26 shows stereo images of a monitor (focal length $12mm$) and in more detail the finger–monitor relation (focal length $69mm$). For this application the control problem is rather complicated. Actually, the goal situation is an ordered sequence of intermediate goal situations which must be reached step by step. Along the course of moving between intermediate goal situations one must keep or reach a further type of goal situation. This means, the measurement vector describing a situation must be partitioned into two subvectors, the first one consisting of attributes which should be kept constant and the second one consisting of attributes which must change systematically.

Fig. 4.26. Stereo images of a monitor, and detailed finger–monitor relation.

For specifying criteria under which the goal situations are reached it is advantageous to *visually demonstrate* these situations in the experimental designing phase. The control cycles for approaching and assembling an effector to a target object are running as long as the deviation between current situation and goal situation is larger than a certain threshold. However, the value for this parameter must be specified in terms of pixels which is inconvenient for system users. Unfortunately, in complicated applications even a vector of threshold values must be specified. To simplify this kind of user interaction it makes sense to manually arrange certain goal situations prior to the servoing cycles and take images.

These images are analyzed with the purpose of automatically extracting the goal situations and furthermore determining relevant thresholds which describe acceptable deviations. For example, for servoing the finger we must specify in terms of pixels the permissible tolerance for the orthogonality to the surface and for the distance from the surface. Actually, these tolerances are a priori known in the euclidean 3D space but must be determined in the images. Figure 4.27 shows in the first and second image exemplary the tolerance concerning orthogonality and distance and in the third and fourth image non-acceptable deviations. For determining the acceptable variances in both parameters a simple image subtraction and a detailed analysis of the subtraction area is useful.

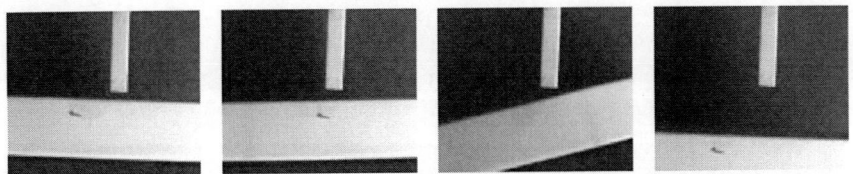

Fig. 4.27. Acceptable and non-acceptable finger–monitor relations.

Keep–Moving Behavior and Keep–Relation Behavior

The specific sub-task, *i.e.* moving an effector around the monitor front and keeping an orthogonal relation and a certain distance, can be solved by *combining two assembled behaviors*. The so-called *keep–moving behavior* is responsible for moving the effector through a set of intermediate positions which approximate the monitor shape coarsly. The so-called *keep–relation behavior* is responsible for keeping the effector in the desired relation to the current part of the surface. The keep–moving behavior strives for moving along an exact rectangle but is modified slightly by the keep–relation behavior. For the keep–moving behavior four intermediate subgoals are defined which are the four corners of the monitor front.

The head-cameras are used for taking stereo images each of which containing the whole monitor front and the gripper finger. In both images we extract the four (virtual) corner points of the monitor by applying approaches of boundary extraction as presented in Chapter 2. By combining the corresponding 2D coordinates between the stereo images we obtain four 4D vectors which represent the intermediate goal positions in the stereo images, *i.e.* we must pass successively four desired measurement vectors $Q_1^*, Q_2^*, Q_3^*, Q_4^*$. The variable state vector $S^v(t)$ is defined as the 3D coordinate vector $P(t)$ of the finger tip, and the current measurement vector $Q(t)$ represents its position in the stereo images. The pseudo inverse $\left(J^f\right)^\dagger (S^v(t))$ of the Jacobian is taken from equation (4.13). The control function for approaching the desired measurement vectors $Q_i^*, i \in \{1, 2, 3, 4\}$, is as follows.

$$C(t) := \begin{cases} s \cdot \dfrac{\left(J^f\right)^\dagger (S^v(t)) \cdot (Q_i^* - Q(t))}{\| \left(J^f\right)^\dagger (S^v(t)) \cdot (Q_i^* - Q(t)) \|} & : \quad \| Q_i^* - Q(t) \| > \eta_1 \\ \qquad\qquad 0 & : \qquad else \end{cases} \quad (4.18)$$

In the application phase parameter i is running from 1 to 4, *i.e.* as soon as Q_i^* is passed taking threshold η_1 into account then the behavior is striving for Q_{i+1}^*. Due to the normalization involved in the control function an increment vector of constant length is computed. This makes sense, because in the inspection sub-task a camera movement with constant velocity is favourable.

The keep–relation behavior is responsible for keeping the finger in an orthogonal orientation near to the current part of the monitor front. For taking images from the situations at a high resolution (see Figure 4.27) the hand-camera CA_3 is used. Similar to the grasping sub-task (treated in Subsection 4.3.4) a rotational and/or a translational movement takes place if the current situation is non-acceptable. For rotational servoing simply histograms of edge orientations can be used to distinguish between acceptable and non-acceptable angles between finger and object surface. Coming back to the role of visual demonstration it is necessary to acquire *three classes of histograms* prior to the servoing cycles. One class consisting of acceptable relations and two other classes representing non-acceptable relations with the distinction of clockwise or counter-clockwise deviation from orthogonality. Based on this, a certain angle between finger and object surface is classified during the servoing cycles using its edge orientation histogram.[21]

For example, a GBF neural network can be used in which a collection of hidden nodes represents the three manifolds of histograms and an output node computes an evidence value indicating the relevant class, *e.g.* value near to 0 for acceptable relations and values near to 1 or -1 for non-acceptable clockwise or counter-clockwise deviation. As usual, the hidden nodes are created on the basis of the *k-means clustering algorithm* and the link weights to the output node are determined by the *pseudo inverse technique*. The control function for the rotation task is similar to equation (4.10) with the distinction that a measure of distance between current and desired measurement vectors (*i.e.* edge orientation histograms) is computed by the RBF network. For translating the finger to reach and then keep a certain distance to the monitor a strategy similar to the grasping approaches can be used (see Subsection 4.3.4).

Task-Specific Module for Continual Handling of an Effector

The cooperation between the keep–moving behavior and the keep–relation behavior is according to the *principle of alternation*. The keep–moving behavior should approach step by step the four corners of the monitor, *i.e.* in each iteration of its control cycles a small increment towards the next monitor corner must be determined. Additionally, in each iteration the second control cycle of the keep–relation behavior must bring the effector into the desired relation to current part of the monitor front. As soon as this relation is reached, the next iteration of the keep–moving control cycle comes into play, and so on. A generic module of type MB_2 would be appropriate for treating two goals in combination. However, concerning the keep–moving behavior we have defined four major subgoals, *i.e.* passing the four corners of

---

[21] The strength of applying the learning process to the raw histogram data is that the network can generalize from a large amount of data. However, if data compression would be done prior to learning (*e.g.* computing symbolic values from the histograms) then quantization or generalization errors are unavoidable.

the monitor front. In consequence of this, the specific sub-task of inspecting or manipulating a monitor front is solved by a sequential combination of the relevant generic module MB_2, and this defines the task-specific module MT_7.

$$MT_7 \quad := \quad (\ MB_2 \quad \& \quad MB_2 \quad \& \quad MB_2 \quad \& \quad MB_2 \ ) \qquad (4.19)$$

4.3.7 Monitoring the Task-Solving Process

For the exemplary high-level task we have worked out the following sub-tasks which must be executed in sequential order: localizing a target object on the domestic area (MT_1), determining and reconstructing obstacle objects (MT_2), approaching, grasping, and clearing away obstacle objects on the parking area (MT_6), approaching, grasping, and carrying the target object to the inspection area (MT_4, MT_5), inspecting and/or manipulating the target object (MT_7).

Based on the definitions of task-specific modules in the previous subsections, we can introduce a task-specific module MT_8 which is used simply for a brief denotation.

$$MT_8 \quad := \quad (\ MT_1 \quad \& \quad MT_2 \quad \& \quad MT_6 \quad \& $$
$$MT_4 \quad \& \quad MT_5 \quad \& \quad MT_7 \ ) \qquad (4.20)$$

Designing Aspects for Monitoring the Task-Solving Process

The overall process, implemented by module MT_8, must be supervised with generic monitors of the types MM_1, MM_2, and MM_3. The *time monitor* MM_1 must check whether the sub-tasks are solved during the periods of time, which are prescribed from the specification of the overall task. It is a matter of the experimental designing phase to implement task-specific modules such that the time constraints can be met in the application phase normally. The *situation monitor* MM_2 must check, after completion of a certain sub-task, whether an environmental situation has been arranged or topical information has been contributed which is needed in successive sub-tasks. For example, in the case that the target object can not be localized on the domestic area, the successive sub-tasks are meaningless and therefore the monitor must interrupt the system.

The *exception monitor* MM_3 must observe the whole environmental scene continually during the task-solving process. This is necessary for reacting appropriately in case of unexpected events. For example, during the application phase the monitor should detect a situation in which a person or an object enters the environment inadmissibly. The problem is to distinguish between goal-oriented events and unexpected events. A rudimentary strategy may work as follows. The environmental scene is subdivided into a field in which the robot arm is working for solving a certain sub-task and the complementary field. This subdivision is changing for each sub-task, *e.g.* during the sub-task of localizing a target object the domestic area is occupied. The ceiling-camera CA_1 is used for detecting events in fields in which the robot arm is not involved, *i.e.* an anexpected event must have occurred.

In the experimental designing phase one must determine for each sub-task the specific field in which the robot arm is working. More concretely, one must determine for each sub-task the specific image area, in which something should happen, and the complementary area, in which nothing should happen. Simple difference image techniques can be applied for detecting gray value changes, but they must be restricted to the areas which are supposed to be constant for the current sub-task. However, if a significant change happens in the application phase nevertheless, then an unexpected event must have occurred and the monitor may interrupt the system.[22]

Task-Specific Module for Monitoring the Task-Solving Process

In a task-specific module MT_9 the generic monitors of the types MM_1, MM_2, and MM_3 must work simultaneously.

$$MT_9 \quad := \quad (\; MM_1 \quad | \quad MM_2 \quad | \quad MM_3 \;) \qquad (4.21)$$

The process of module MT_8 must be supervised by the process of module MT_8 continually. For this simultaneous execution we introduce the final task-specific module MT_{10}.

$$MT_{10} \quad := \quad (\; MT_8 \quad | \quad MT_9 \;) \qquad (4.22)$$

4.3.8 Overall Task-Specific Configuration of Modules

For solving the exemplary high-level task we introduced a series of task-specific modules, *i.e.* $MT_1, \cdots, MT_{10}$. They are defined as sequential and/or parallel configurations of generic modules taken from the repository (presented in Section 4.2). As a summary, Figure 4.28 shows the overall configuration of modules which defines the autonomous camera-equipped robot system for solving the exemplary task.

According to the vertical organization the generic modules MI_1, MI_2, MI_3, MB_3 are based on vector fields from the deliberate layer, and the generic modules MB_1, MB_2, MB_3 make use of visual feedback at the reactive layer. Generic module MB_3 integrates deliberate and reactive processing (which is also done in generic modules MB_5 and MB_6, presented in Appendix 2). Also the task-specific modules combine deliberate and reactive processing, which is obvious due to the sequential and/or parallel combination of the generic modules. Two effectors contribute to the exemplary task, *i.e.* the

---

[22] Of course, this strategy is not able to detect unexpected events in the field in which the robot arm is working currently. More sophisticated approaches are necessary for treating problems like these.

$$MT_{10} := \left(/ \middle| MT_8 := \left| \begin{array}{l} MT_1 := (MI_1 \ \& \ MI_2 \ \& \ MI_3 \ \& \ MI_2) \\ \& \\ MT_2 := (MI_3 \ \& \ MI_2) \\ \& \\ MT_6 := (MT_3 \ \& \ MT_4 \ \& \ MT_5)^* \\ \& \\ MT_4 := (MB_3 \ \& \ MB_1 \ \& \ MB_1 \ \& \\ \qquad\qquad MB_1 \ \& \ MB_1 \ \& \ MI_2) \\ \& \\ MT_5 := (MI_2 \ | \ MI_2) \ \& \ MB_1 \ \& \ (MI_2 \ | \ MI_2) \\ \& \\ MT_7 := (MB_2 \ \& \ MB_2 \ \& \ MB_2 \ \& \ MB_2) \end{array} \right. \\ \\ MT_9 := (MM_1 \ | \ MM_2 \ | \ MM_3) \right)$$

Fig. 4.28. Configuration of task-specific modules based on generic modules for solving the exemplary high-level task.

robot arm and the robot head. Therefore, two types of state vectors are involved which results in a horizontal organization consisting of two columns (see Figure 4.6).

We configured the system under the guiding line of minimalism principles. A minimal set a behavioral and instructional modules is involved and each one is responsible for executing a specific sub-task. Every module is essential, and if one of them fails, then the exemplary task cannot be solved. Three types of specific modules are used for supervising the task-solving process, *i.e.* time, situation, and exception monitor. A high-level task must be solved completely, *i.e.* partial solutions which may be caused by module failures are not acceptable. To design robust, task-solving systems, which are tolerant against module failures, one must include redundant modules. It is the responsibility of the monitors to recognize erroneous module behaviors and to bring alternative modules into application. Our work does not treat this aspect in more detail.

In the next section we introduce basic mechanisms for camera–robot coordination which have not been treated so far.

4.4 Basic Mechanisms for Camera–Robot Coordination

For treating the exemplary task we take advantage of certain constant features which characterize the relations between various cameras and the robot arm. For approximating the relevant relations we make use of the *agility of the robot arm* and learn from systematic hand movements. That is, in the experimental designing phase a set of samples, which consist of corresponding robot and image coordinates, is memorized and relevant relations are approximated from those. For example, this strategy has been applied for determining the relations between various coordinate systems in our double-eye-on-hand system, *i.e.* robot hand and the hand-cameras CA_2 and CA_3. We refer to the diploma thesis of Kunze for theoretical foundations and experimental results [94].[23] In this section, we concentrate on acquiring the relations involved in our double-eye-off-hand system, *i.e.* robot hand and the head-cameras CA_4 and CA_5. The description also includes the optical axes and the fields of view. All of these features are represented relative to the static coordinate system of the robot arm and can be changed by the degrees of freedom of the head system. For the automatic alignment of the head-camera system, including the movement of the robot vehicle to the optimal place, it would be necessary to design task-specific modules based on the generic modules presented in Section 4.2. However, this is beyond the scope of the chapter and therefore we assume that the system designer does the work manually.

4.4.1 Camera–Manipulator Relation for One-Step Control

The relevant modality of the head-camera–manipulator relation depends on specific constraints which are inherent in the characteristics of a task. In this subsection we consider the sub-task of moving the robot hand to a certain position which is located in the field of view of the head-cameras, but starting from a position outside the field of view. For example, in our exemplary task we treated the sub-task of carrying an object from the domestic area to the parking area, but only the latter area was located in the field of view. For such applications visual feedback control is not possible, because the robot hand is not visible in the early phase, *i.e.* we can not extract a current measurement vector $Q(t)$ from the images. Therefore, the only way is to extract the desired measurement vector Q^*, reconstruct the 3D position as accurate as possible, and move the robot hand in *one step* to the determined position. In this section the reconstruction function is approximated nonlineary by GBF networks (see Subsection 3.2.2 for foundations of GBF networks). For various configurations we compute the *reconstruction errors* in order to obtain a reasonable network structure which would be appropriate for reaching a certain degree of accuracy.

---

[23] Diploma thesis (in german) supervised by the author of this book.

Relationship between Image and Effector Coordinate Systems

By taking stereo images with the head-cameras and detecting the target place in the two images, we obtain two two-dimensional positions (*i.e.* two 2D vectors). The two positions are defined in the coordinate systems of the two cameras and are combined in a single vector (*i.e.* 4D vector). On the other hand, the robot hand moves within a 3D working space, which is defined in the basis coordinate system of the robot arm. The position of the virtual gripper point is a 3D vector which is located in the middle between the finger tips of the robot hand. Hence, we need a function for transforming the target positions from the image coordinate systems of the cameras to the cartesian coordinate system of the robot arm (*i.e.* transforming 4D vectors into 3D vectors).

Traditionally, this function is based on principles of stereo triangulation by taking intrinsic parameters (of the camera) and extrinsic parameters (describing the camera–robot relationship) into account [52]. Opposed to that, we use GBF networks to learn the mapping from stereo image coordinates into coordinates of a robot manipulator. There are *three good reasons* for this approach. First, the intrinsic and extrinsic parameters are unnecessary and therefore are not computed explicitly. The coordinate mapping from stereo images to the robot manipulator is determined in a direct way without intermediate results. Second, usual approaches of camera calibration assume certain camera models which must be known formally in advance, e.g. perspective projection und radial distortion. Instead of that, the learning of GBF networks takes place without any a priori model and can approximate any continuous projection function. Third, by varying the number and the parametrization of the GBFs during the training phase, the accuracy of the function approximation can be controlled as desired. For example, a coarse approximation would be acceptable (leading to a minimal description length) in applications of continual perception-action cycles.

Acquiring Training Samples by Controlled Effector Movements

The procedure for determining the camera–robot coordination is as follows. We make use of training samples for learning a GBF network. First, the set of GBFs is configured, and second, the combination factors of the GBFs are computed. We configure the set of GBFs by simply *selecting certain elements* from the whole set of training samples and using the input parts (4D vectors) of the selected samples to define the centers of the GBFs. The combination factors for the GBFs are computed with the pseudo inverse technique, which results in *least square errors* between pre-specified and computed output values.

The prerequisite for running the learning procedure is the existence of training samples. In order to obtain them, we take full advantage of the agility of the robot arm. The *hand effector moves* in the working space systematically and stops at equidistant places. For each place we record the 3D position of the virtual gripper point of the robot hand which is equal to the position of the place supplied to the control unit of the robot arm, *i.e.* the tool center point. Furthermore, at each stopping place a *correlation-based recognition algorithm* detects the gripper tip in the stereo images (see Figure 4.29) and the two two-dimensional positions are combined to a 4D vector. All pairs of 4D–3D vectors are used as training samples for the desired camera–robot coordination.

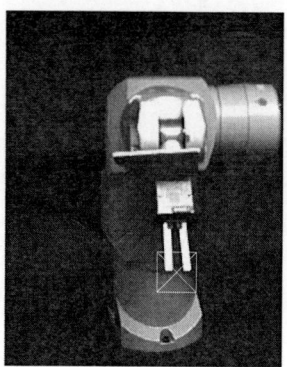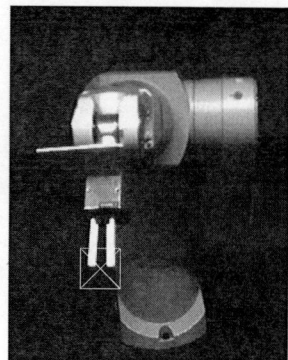

Fig. 4.29. The stereo images show the robot hand with parallel jaw fingers. A correlation-based recognition algorithm has been used to localize the virtual gripper point. This is illustrated by a white square including two intersecting diagonals.

The strategy of using the robot arm itself for determining the head-camera–manipulator relation is advantageous in several aspects. First, we don't need an artificial calibration object. Second, samples can be taken both from the surface and within the working space. Third, the number of samples for approximating the function is variable due to steerable distances between the stopping places. Fourth, the head-camera–manipulator relation is computed relative to the basis coordinate system of the robot arm directly, which is the relevant coordinate system for controlling the robot hand.

Experiments to the Estimation of the Camera–Robot Relationship

Based on image coordinates of the virtual gripper point, the GBF network has to estimate its 3D position in the basis coordinate system of the robot arm. On average the 3D position error should be as low as possible. The main question of interest is, how many GBFs and which extents are needed to obtain a certain quality for the camera–robot coordination. In order to answer this question, *four experiments* have been carried out. In the first and second experiment, we applied two different numbers of GBFs exemplary. The third experiment shows the effect of doubling the image resolution. Finally, the fourth experiment takes special care for training the combination weights of the GBFs. In all four experiments, we systematic increase the GBF extent and evaluate the mean position error.

We take training samples for each experiment. The working space of the robot hand is cube–shaped of $300mm$ side length. The GBFs are spread over a sup-space of 4D vectors in correspondence to certain stopping places of the robot hand. That is, the 4D image coordinates (resulting from the virtual gripper point at a certain stopping place) are used for defining the center of a Gaussian basis function. The following experiments differ with regard to the size and the usage of the *training samples*. The application of the resulting GBF networks is based on *testing samples* which consist of input–output pairs from the same working space as above. For generating the testing samples the robot hand moves in discrete steps of $20mm$ and it is assured that training and testing samples differ for the most part, *i.e.* have only a small number of elements in common.

In the *first experiment*, the robot hand moved in discrete steps of $50mm$ through the working space which result in $7 \times 7 \times 7 = 343$ training samples. Every second sample is used for defining a GBF ($4 \times 4 \times 4 = 64$ GBFs) and all training samples for computing the combination weights of the GBFs. The image resolution is set to 256×256 pixel. Figure 4.30 shows in curve (a) the course of mean position error (of the virtual gripper point) for a systematic increase of the Gaussian extent. As the GBFs become more and more overlapped the function approximation improves, and the mean position error decreases to a value of about $2.2mm$.

The *second experiment* differs from the first in that the robot hand moved in steps of $25mm$, *i.e.* $13 \times 13 \times 13 = 2197$ training samples. All samples are used for computing the GBF weights, and every second sample for defining a GBF ($7 \times 7 \times 7 = 343$ GBFs). Figure 4.30 shows in curve (b) that the mean position error converges to $1.3mm$.

In the *third experiment* the same configuration has been used as before, but the image resolution was doubled to 512×512 pixels. The accuracy of localizing the finger tips in the images increases, and hence the mean position error of the virtual gripper point reduces once again. Figure 4.30 shows in curve (c) the convergence to error value $1.0mm$.

The *fourth experiment* takes special care of both the training of weights and the testing of the resulting GBF network. Obviously, there is only a one-sided overlap between GBFs at the border of the working space. Hence, the quality of the function approximation can be improved if a specific sub-set of 3D vectors, which is located at the border of the working space, will not be taken into account. In this experiment, the 343 GBFs are spread over the original working space as before, but an inner working space of $250mm$ side length is used for computing combination factors and for testing the GBF network. Figure 4.30 shows in curve (d) that the mean position error decreases to a value of $0.5mm$.

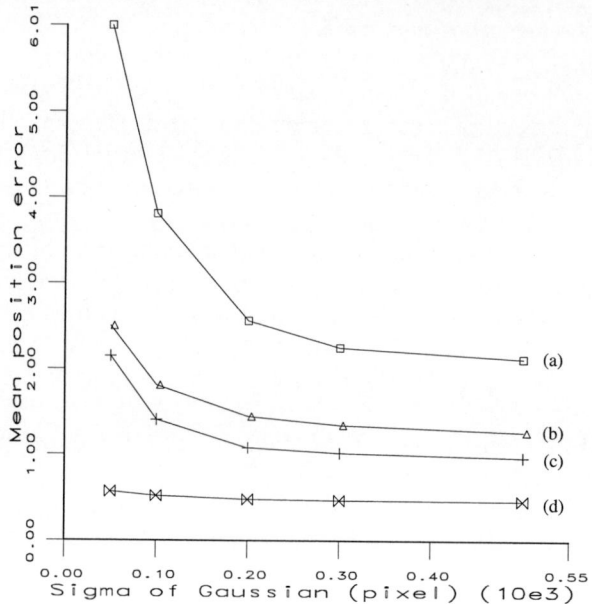

Fig. 4.30. The curves show the mean position error versus the extents of GBFs under four different conditions. (a) Small GBF number, low image resolution. (b) Large GBF number, low image resolution. (c) Large GBF number, high image resolution. (d) Experiment (c) and avoiding approximation errors at working space border. Generally, the error decreases by increasing the Gaussian extent, and the larger the GBF number or the higher the image resolution the smaller the position error.

Conclusions from the Experiments

Based on these experiments, we configure the GBF network such that a desired accuracy for 3D positions can be reached (*e.g.* $\pm 0.5mm$). During the application phase, first the target place must be detected in the stereo images. Second, the two 2D coordinate vectors are put into the GBF network for computing a 3D position. Finally, the robot hand will move to that 3D position which is approximately the position of the target place.

Although the obtainable accuracy is impressive this approach can not react on unexpected events during the movement of the robot hand, because continual visual feedback is not involved. However, there are applications in which unexpected events are excluded, and for these cases the approach would be favourable.

4.4.2 Camera–Manipulator Relation for Multi-step Control

In applications such as putting down an object on a physical target place
(*e.g.* on the parking or inspection area) or steering the hand effector over a
large object (*e.g.* monitor front case), the current position of the effector is
in *close spatial neighborhood to the target place*. In order to avoid undesired
events such as collisions, one must be careful and move the robot hand just
in *small steps under continual visual feedback*, *i.e.* applying a *multi-step* controller. Due to the close spatial neighborhood, it is possible to take relevant
images such that both the current and the desired measurement vectors can
be extracted. This is the precondition for applying procedures of image-based
effector servoing, as worked out in Subsection 4.2.1.

A basic constituent of the servoing procedure is a description of the relationship between displacements of the robot hand and the resulting displacements in the image of a head-camera. A usual approach is to approximate the
projection function, which transforms 3D robot coordinates into 2D image
coordinates, and specify the Jacobian matrix. The Jacobians have already
been applied in Subsections 4.3.4, 4.3.5, and 4.3.6. In this subsection we
specify two variants of projection functions, *i.e.* a *linear and a nonlinear approximation*, and determine the Jacobians thereof. The training samples of
corresponding 3D points and 2D points are determined according to the approach mentioned in the previous subsection, *i.e. tracking controlled gripper
movements*.

Jacobian for a Linear Approximation of the Projection Function

A *perspective projection matrix* $\mathcal{Z}$ lineary approximates the relation between
the manipulator coordinate system and the image coordinate system of a
head-camera.

$$\mathcal{Z} := \begin{pmatrix} \mathcal{Z}_1^v \\ \mathcal{Z}_2^v \\ \mathcal{Z}_3^v \end{pmatrix} \; ; \; with \quad \begin{matrix} \mathcal{Z}_1^v := (z_{11}, z_{12}, z_{13}, z_{14}) \\ \mathcal{Z}_2^v := (z_{21}, z_{22}, z_{23}, z_{24}) \\ \mathcal{Z}_3^v := (z_{31}, z_{32}, z_{33}, z_{34}) \end{matrix} \qquad (4.23)$$

The usage of the projection matrix is specified within the following
context. Given a point in homogeneous manipulator coordinates $P :=
(X, Y, Z, 1)^T$, the position in homogeneous image coordinates $p := (x, y, 1)^T$
can be obtained by solving

$$p := f^{li}(P) := \begin{pmatrix} f_1^{li}(P) \\ f_2^{li}(P) \\ f_3^{li}(P) \end{pmatrix} := \frac{1}{\xi} \cdot \mathcal{Z} \cdot P \; ; \; with \; \xi := \mathcal{Z}_3^v \cdot P \qquad (4.24)$$

According to this, the matrix $\mathcal{Z}$ is determined with simple linear methods
by taking the training samples of corresponding 3D points and 2D points into
account [52, pp. 55-58]. The scalar parameters z_{ij} represent a combination
of extrinsic and intrinsic camera parameters which we leave implicit. The
specific definition of the normalizing factor ξ in equation (4.24) guarantees

that function $f_3^{li}(P)$ is constant 1, *i.e.* the homogeneous image coordinates of position p are given in normalized form.

Next, we describe how a certain change in manipulator coordinates affects a change in image coordinates. The Jacobian J^f for the transformation f^{li} in equation (4.24) is computed as follows.

$$J^f(P) := \begin{pmatrix} \frac{\partial f_1^{li}}{\partial X}(P) & \frac{\partial f_1^{li}}{\partial Y}(P) & \frac{\partial f_1^{li}}{\partial Z}(P) \\ \frac{\partial f_2^{li}}{\partial X}(P) & \frac{\partial f_2^{li}}{\partial Y}(P) & \frac{\partial f_2^{li}}{\partial Z}(P) \end{pmatrix}$$

$$:= \begin{pmatrix} \frac{z_{11} \cdot \mathcal{Z}_3^v \cdot P - z_{31} \cdot \mathcal{Z}_1^v \cdot P}{(\mathcal{Z}_3^v \cdot P) \cdot (\mathcal{Z}_3^v \cdot P)} & \cdots \\ \vdots & \ddots \end{pmatrix} \tag{4.25}$$

These computations must be executed for both head-cameras CA_2 and CA_3 which result in two perspective projection matrices $\mathcal{Z}_2$ and $\mathcal{Z}_3$ and two Jacobians J_2^f and J_3^f.

Jacobian for a Nonlinear Approximation of the Projection Function

Instead of using a projection matrix we can also take a GBF network for approximating (nonlinear) the relation between the manipulator coordinate system and the image coordinate system of a head-camera. The definitions of the functions f_1^{li} and f_2^{li} in equation (4.24) must be redefined as a weighted sum of Gaussians, respectively.

$$f_j^{nl}(P) := \sum_{i=1}^{I} w_{ij} \cdot f_i^{Gs}(P) \; ; \; j \in \{1, 2\} \tag{4.26}$$

$$\text{with } f_i^{Gs}(P) := exp\left(-\frac{\|P - \overline{P_i}\|^2}{2 \cdot \sigma_i^2}\right)$$

These equations represent a GBF network with a two-dimensional output. The centers $\overline{P_i}$ and extents σ_i, for $i \in \{1, \cdots, I\}$, are obtained by usual approaches of GBF network learning.

The Jacobian J^f for the redefined transformation f^{nl} is as follows.

$$J^f(P) := \begin{pmatrix} \frac{\partial f_1^{nl}}{\partial X}(P) & \frac{\partial f_1^{nl}}{\partial Y}(P) & \frac{\partial f_1^{nl}}{\partial Z}(P) \\ \frac{\partial f_2^{nl}}{\partial X}(P) & \frac{\partial f_2^{nl}}{\partial Y}(P) & \frac{\partial f_2^{nl}}{\partial Z}(P) \end{pmatrix}$$

$$:= \begin{pmatrix} -\frac{1}{\sigma_i^2} \cdot \sum_{i=1}^{I} w_{i1} \cdot f_i^{Gs}(P) \cdot (X - \overline{X_i}) & \cdots \\ \vdots & \ddots \end{pmatrix} \tag{4.27}$$

These computations must be executed for both head-cameras CA_2 and CA_3 which result in two GBF networks and two Jacobians J_2^f and J_3^f. For determining appropriate translations of the robot hand we must combine the two Jacobians, compute the pseudo inverse, and apply the resulting matrix to the difference vector between desired and current measurement vectors in the images. This procedure has already been applied in Subsection 4.3.5 (see equations (4.12), (4.13), (4.14), and (4.15)).

Dealing with Inaccurate Head-Camera–Manipulator Relations

In Subsection 4.4.1, we showed exemplary that GBF networks can approximate the transformation between coordinates in stereo images and coordinates in the robot arm up to an impressive degree of accuracy. However, the accuracy decreases considerable if the relation between head-camera and robot arm will be changed physically by accident. Instead of approximating the relation again, we can alternatively work with the inaccurate approximation of head-camera–manipulator relation and make use of the *servoing mechanism*. The following experiments are executed in order to confirm this strategy.

Servoing Experiments under Inaccurate Head-Camera–Manipulator Relations

The spatial distance between the center of the head-camera system and the center of the robot arm is about $1500mm$, the focal length of the two head-cameras has been steered to $12mm$, respectively. The working space of the robot hand is a cube of sidelength $400mm$. Projection matrices are used for approximating the projection function between robot and image coordinates. Three different approximations are considered which are based on different densities of the training samples. Concretely, for the *first approximation* the stopping places of the robot hand were at distances of $100mm$ which yielded 125 training samples, for the *second approximation* the hand stopped every $200mm$ which yielded 27 training samples, and for the *third approximation* the hand stopped every $400mm$ (*i.e.* only at the corners of the working space) which yielded 8 training samples. For each approximation we compute two Jacobians, respectively for each head-camera and combine them according to equation (4.13). In all experiments the gripper must start at a corner and is supposed to be servoed to the center of working space by applying the control function in equation (4.15).

For a servoing factor $s := 0.5$ it turns out that at most 10 cycle iterations are necessary until convergence. After convergence we make measurements of the deviation from the 3D center point of the working space. First, the servoing procedure is applied under the use of the three mentioned projection matrices. The result is that the final deviation from the goal position is at most $5mm$ with no direct correlation to the density of the training samples (*i.e.*, the accuracy of the initial approximation). According to this, it is sufficient to use just eight corners of the working space for the approximation of the head-camera–manipulator relation. Second, the servoing procedure is applied after changing certain degrees of freedom of the robot head, respectively, and thus *simulating various accidents*. Changing the head position in a circle of radius $100mm$, or changing pan or tilt DOF within angle interval of $10°$ yield deviations from goal position of at most $25mm$. The errors occur mainly due to the restricted image resolution of 256×256 pixels. According to these results, we can conclude that a multi-step control procedure is able to deal with rough approximations and accidental changes of the head-camera–manipulator relation.

Handling Inaccurate Relations by Servoing Mechanisms

The experiments proved, that image-based effector servoing plays a fundamental role in the application phase of the process of solving a high-level, deliberate task (see Section 4.3). In addition to this, the servoing mechanism can also *support the experimental designing phase* which precedes the application phase. For example, for certain tasks of active vision one must determine additional features of the camera and/or the relation between robot and camera, *i.e.* in addition to the coordinate transformations treated in Subsections 4.4.1 and 4.4.2. Specifically, these additional features may comprise the optical axis and the field of view of the head-cameras relative to the basis coordinate system of the robot arm.

4.4.3 Hand Servoing for Determining the Optical Axis

The optical axis plays a fundamental role for supporting various techniques of image processing. For example, the robot arm may carry an object into the field of view of a head-camera, then *approach the object along the optical axis* to the camera, and finally inspect the object in detail.[24] Both sub-tasks of approaching the object to the camera and detailed object inspection can be controlled, respectively, carried out by techniques of image processing which are concentrated in an area around the image center.

Servoing Strategy for Estimating the Optical Camera Axis

For *estimating the optical axis* of a head-camera relative to the basis coordinate system of the robot arm we present a strategy which is based on image-based hand effector servoing. The virtual gripper point of the robot hand is servoed to two distinct points located on the optical axis. It is assumed that all points located on this axis are projected to the image center approximately. Accordingly, we must servo the robot hand such that the two-dimensional projection of the virtual gripper point approaches the image center. In the goal situation the 3D position of the virtual gripper point

---

[24] The object resolution in the images can also be increased by changing the focal length of the head-camera lens (see Subsection 4.3.6). Therefore, a cooperative work of changing the external DOF of the robot arm and the internal DOF of the head-camera may reveal optimal viewing conditions.

(which is the known tool center point in the (X, Y, Z) manipulator coordinate system) is taken as a point on the optical axis. Two planes are specified which are parallel to the (Y, Z) plane with constant offsets X_1 and X_2 on the X-axis. The movement of the virtual gripper point is restricted just to these planes (see Figure 4.31).

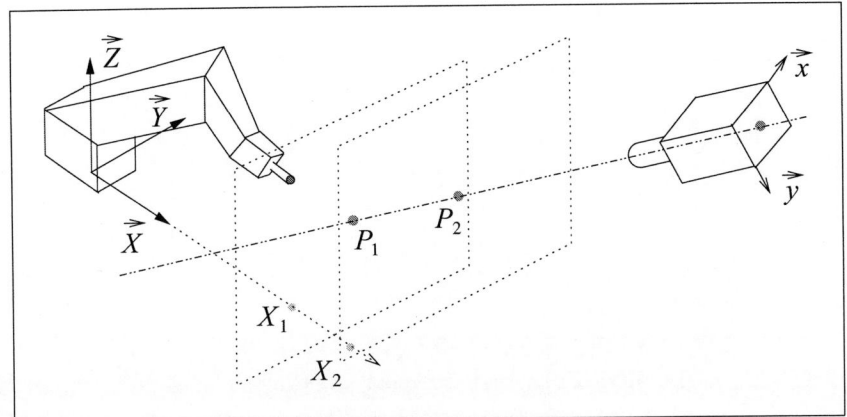

Fig. 4.31. Determining the optical axis of a head-camera.

In image-based effector servoing the deviation between a current situation and a goal situation is specified in image coordinates. In order to transform a desired change from image coordinates back to manipulator coordinates the inverse or pseudo inverse of the Jacobian is computed. For this sub-task the generic definition of the Jacobian J^f (according to equation (4.25) or equation (4.27)) can be restricted to the second and third columns, because the coordinates on the X-axis are fixed. Therefore, the inverse of the quadratic Jacobian matrix is computed, *i.e.* $\left(J^f\right)^{\dagger}(P) := \left(J^f\right)^{-1}(P)$.

Control Function for the Servoing Mechanism

The current measurement vector $Q(t)$ is defined as the 2D image location of the virtual gripper point and the desired measurement vector Q^* as the image center point. The variable state vector $S^v(t)$ consists of the two variable coordinates of the tool center point in the selected plane (X_1, Y, Z) or (X_2, Y, Z). With these redefinitions of the Jacobian we can apply the control function which has already been presented in equation (4.15). The hand position is changed by a non-null vector $C(t)$ if the desired and the current position in the image deviate more than a threshold η_1. According to our strategy, first the virtual gripper point is servoed to the intersection point P_1 of the unknown optical axis with the plane (X_1, Y, Z), and second to the intersection point P_2 with plane (X_2, Y, Z). The two resulting positions of the virtual gripper point specify the axis which is represented in the basis co-ordinate system of the robot arm. Figure 4.32 shows for the hand servoing on one plane the succession of the virtual gripper point extracted in the image, and the final point is located at the image center (servoing factor $s := 0.3$).

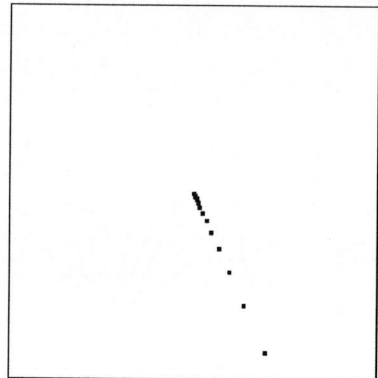

Fig. 4.32. Course of detected virtual gripper point in the image.

4.4.4 Determining the Field of Sharp View

Image-based hand effector servoing is also a means for constructing the *field of sharp view* of a head-camera which can be approximated as a *truncated pyramid* with top and bottom rectangles normal to the optical axis (see Figure 4.33). The top rectangle is small and near to the camera, the bottom rectangle is larger and at a greater distance from the camera.

Servoing Strategy for Determining the Depth Range of Sharp View

For determining the *depth range* of sharp view the virtual gripper point is servoed along the optical axis and the sharpness of the depicted finger tips is evaluated. As the finger tips are located within an area around the image center, we can extract a relevant rectangular patch easily and compute the sharpness in it (see Subsection 4.3.6 for a possible definition of sharpness measurements). Figure 4.34 shows these measurements for a head-camera with focal length $69mm$. The robot hand starts at a distance of $1030mm$ to the camera and approaches to $610mm$ with stopping places every $30mm$ (this gives 15 measurements). We specify a threshold value Q^* for the measurements $Q(t)$ for defining the acceptable level of sharpness. In Figure 4.34 four measurements surpass the threshold, *i.e.* numbers $8, 9, 10, 11$, which means that the depth range of sharpness is about $90mm$, reaching from $700mm$ to $790mm$ distances from the camera.

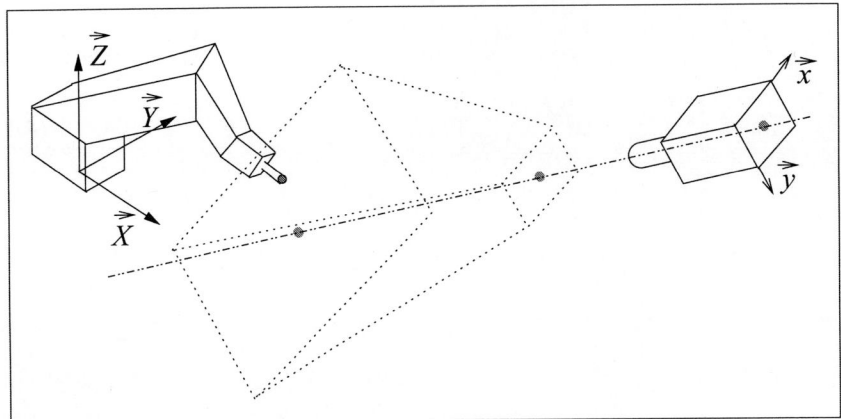

Fig. 4.33. Pyramid field of view and sharpness.

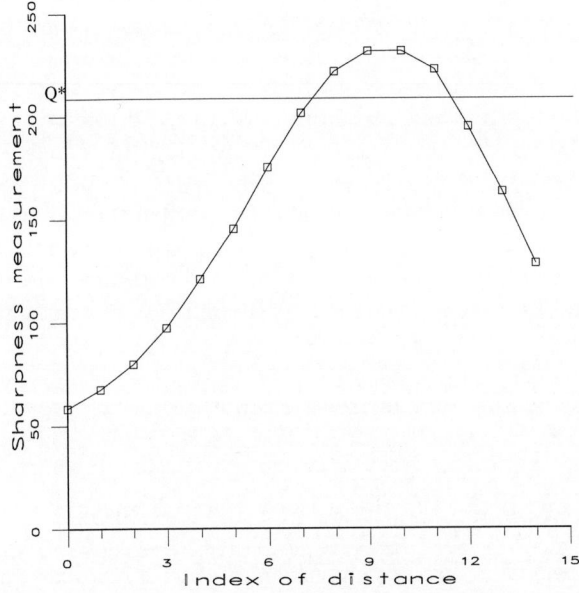

Fig. 4.34. Sharpness measurements in the image section containing the finger tips; course for approaching the fingers along the optical axis of a head-camera.

Control Functions for the Servoing Mechanism

The control procedure consists of two stages, first *reaching the sharp field*, and second *moving through it.*

$$C_1(t) := \begin{cases} s & : & (Q^* - Q(t)) > 0 \\ 0 & : & else \end{cases} \tag{4.28}$$

$$C_2(t) := \begin{cases} s & : & (Q^* - Q(t)) < 0 \\ 0 & : & else \end{cases} \tag{4.29}$$

The variable state vector $S^v(t)$ is just a scalar defining the position of the virtual gripper point on the optical axis and the control vector $C(t)$ is constant scalar (*e.g.* $s := 30mm$). As a result of this procedure, we obtain the top and bottom point on the optical axis which characterize the depth range of sharpness. The *width and height* of the field of sharp view must be determined at these two points which are incident to the top and bottom rectangle of the truncated pyramid. Once again, the agility of the manipulator comes into play to determine the rectangle corners. First, the virtual gripper point must servoed on the top plane and, second, on the bottom plane. Sequentially, the gripper should reach those four 3D positions for which the virtual gripper point is projected onto one of the image corners. The control schema is equal to the one for determining the optical axis with redefined measurement vectors and control vectors. By repeating the procedure for both planes, we obtain the eight corners of the truncated pyramid. For example, using quadratic images from the our head-camera (focal length $69mm$) the sidelength of the top rectangle is $80mm$ and of the bottom rectangle $90mm$.

We presented an approach for determining the field of view of a camera based on the optical axis. Several other applications of optical axes are conceivable. For example, the head position can be determined relative to the coordinate system of the robot arm; by changing the pan and/or tilt DOF of the robot head into two or more discrete states, determining the optical axis, respectively, and intersecting the axes. Although it is interesting, we do not treat this further.

The final Section 4.5 summarizes the chapter.

4.5 Summary and Discussion of the Chapter

This chapter presented basic mechanisms and generic modules which are relevant for designing autonomous camera-equipped robot systems. For solving a specific high-level task we performed the designing phase and thus we developed an exemplary robot vision system. In particular, we did experiments in the task-relevant environment to construct image operators and adaptive procedures, and thus implemented task-specific modules and combinations thereof. Intentionally, a simple high-level task has been chosen in order to clearly present the principles of developing autonomous camera-equipped robot systems. By relying on a repertoire of mechanisms and modules one can develop more sophisticated robot systems for solving more complicated high-level tasks. Generally, the designing/developing phase must treat the following aspects:

- Determine the way of decomposing the high-level task into sub-tasks,
- determine whether to solve sub-tasks with instructional or behavioral modules,
- determine how to integrate/combine deliberate plans and visual feedback control,
- determine how to generate deliberate plans from visual input,
- determine which type of controller is useful for specific sub-tasks,
- determine the consequences of the specific hardware in use (e.g. type of camera objectives, type of gripper fingers) with regard to image analysis and effector tracking,
- determine strategies for arranging cameras such that image processing is simplified (*e.g.* degrees of compatibility, simplicity of appearance manifolds),
- determine strategies of moving cameras and taking images such that the task-relevant information can be extracted at all,
- verify the appropriateness of image processing techniques along with specific parameters,
- construct adaptive procedures such that image processing parameters can be fine-tuned automatically, if needed,
- learn operators for object and/or situation recognition,
- construct goal-directed visual servoing procedures along with specific parameters,
- construct exploration strategies which are based on immediate rewards or punishments instead of explicit goals,
- determine camera–robot relations depending on the intended strategy of using them,
- determine a reasonable time-schedule which must be kept for executing the sub-tasks,
- determine a strategy for reacting appropriately on unexpected events,
- *etc.*

As a result of the designing/developing phase, it is expected to obtain a configuration of appropriate task-specific modules for treating and solving a high-level task during the application phase, autonomously. Despite of the indisputable advantages of the bottom-up designing methodology the task-solving process may fail nevertheless, especially in case of completely new aspects occuring in the application phase. In consequence of that, the designing/developing phase and the application phase must be integrated more thoroughly with the main purpose that learning should take place during the whole process of task-solving and should not be restricted to the designing/developing phase.

5. Summary and Discussion

We presented a paradigm of developing camera-equipped robot systems for high-level Robot Vision tasks. The final chapter summarizes and discusses the work.

5.1 Developing Camera-Equipped Robot Systems

Learning-Based Design and Development of Robot Systems

There are no generally accepted methodologies for building embedded systems. Specifically, autonomous robot systems can not be constructed on the basis of pre-specified world models, because they are hardly available due to imponderables of the environmental world. For designing and developing autonomous camera-equipped robot systems we propose a learning-based methodology. One must demonstrate relevant objects, critical situations, and purposive situation-action pairs in an experimental phase prior to the application phase. Various learning machines are responsible for acquiring image operators and mechanisms of visual feedback control based on supervised experiments in the task-relevant, real environment. This supervisory nature of the experimental phase is essential for treating high-level tasks in the application phase adequately, *i.e.* the behaviors of the application phase should meet requirements like task-relevance, robustness, and time limitation simultaneously.

> Autonomous camera-equipped robot systems must be designed and developed with learning techniques for exploiting task-relevant experiences in the real environment.

Learning Feature Compatibilities and Manifolds

Some well-established scientists of the autonomous robotics community expressed doubts versus this central role of learning. They argued that tremendous numbers of training samples and learning cycles are needed and that

J. Pauli: Learning-Based Robot Vision, LNCS 2048, pp. 255-261, 2001.
© Springer-Verlag Berlin Heidelberg 2001

the overgeneralization/overfitting dilemma makes it difficult to acquire really useful knowledge. However, in this work we introduced some principles which may overcome those typical learning problems to a certain extent. For learning image operators we distinguished clearly between compatibilities and manifolds. A compatibility is related to a feature which is more or less stable under certain processes, *e.g.* certain proprioceptive features of the robot actuator system or the camera system, certain inherent features of rigid scene objects and their relationships. Such features help to recognize certain facets of the task-solving process again and again despite of other changes, *e.g.* recognizing the target position during a robot navigation process. Complementary to this, manifolds are significant variations of image features which originate under the task-relevant change of the spatial relation between robot effector, cameras, and/or environmental objects. The potential of manifolds is to represent systematic changes of features which are the foundation for steering and monitoring the progress of a task-solving process. For solving a high-level robotic task both types of features are needed and are applied in a complementary fashion.

> The matter of systematic experimentations in the real environment is to determine realistic variations of certain features, *i.e.* learn feature compatibilities and manifolds.

Balance between Compatibilities and Manifolds

Mathematically, a compatibility is represented by the mean and the variance of the feature, or in case of a feature vector by the mean vector and the covariance matrix. The underlying assumption is a multi-dimensional Gaussian distribution of the deviations from the mean. In particular, the description length of a compatibility is small. Instead of that, a manifold is much more complicated, and therefore, the necessary approximation function goes beyond a simple Gaussian. In order to reduce the effort of application during the online phase, the complexity of the manifold should be just as high as necessary for solving the task. Certain compatibilities can be incorporated with the potential of reducing the complexity of related manifolds, *e.g.* by applying Log-polar transformation to images of a rotating camera one reduces the manifold complexity of the resulting image features. The combination of features, which will be involved in a task-solving process, is determined on the basis of evaluations in the experimentation phase. A compromise is needed for conflicting requirements such as task-relevance, robustness, and limited period of time. This compromise is obtained by a balance between degrees of compatibilities on one one hand and complexities of manifolds on the other hand for the two types of features.

> For solving a high-level robotic task, the experimentation phase must result in a balance between two types of features, *i.e.* approximate constant features (compatibilities) versus systematic changing features (manifolds).

Exploiting Degrees of Freedom for Solving Tasks

A high-level robotic task leaves some degrees of freedom or even accepts different strategies for a solution, *e.g.* different trajectories of robotic object grasping, or different camera poses for monitoring a manipulation task. The latter example belongs to the paradigm of *Active Vision*. The degrees of freedom must be exploited with the purpose of reducing the complexities of features compatibilities and manifolds. The process of system designing is driven to a certain extent by minimal degrees of compatibilities, which are needed for the applicability of a priori image operators, and also driven by maximal complexities of manifolds, which are accepted for applying learned image operators within time limits. In the experimentation phase it is up to the system designer to determine relations between the camera system and other environmental constituents, such that certain complexities and balances of the compatibilities and manifolds are expected to hold. In the subsequent application phase a first sub-task is to steer the robot actuators with the purpose of automatically arranging relations which are consistent with the pre-specified relations as considered in the experimentation phase.

> Camera steering modules are responsible for arranging camera–environment relations such that certain image operators will be applicable.

Consensus with the Paradigm of Purposive, Animate Vision

Our approach of developing autonomous camera-equipped robot systems is in consensus with the paradigm of *Purposive Vision* [2]. The compatibilities introduced in chapter 2, the learning mechanisms proposed in chapter 3, and the generic modules presented in chapter 4 are applicable to a large spectrum of high-level tasks. The learned compatibilities and manifolds and the generic modules approximate general assumptions underlying the process of embedding the Robot Vision system in the environment. Based on that, an autonomous system can be developed and adapted for an exemplary high-level task. A specific task is solved under general assumptions and by applying and adapting general mechanisms. By incorporating learning mechanisms this methodology goes far beyond the paradigm of Purposive Vision leading to *Animate Vision*. It has been recognized in the paradigm of Animate Vision that learning is necessary to compensate for the world's unpredictability [12].

> The development of autonomous, camera-equipped robot systems follows the paradigm of purposive, animate vision.

General Systems versus General Development Tools

In Subsection 1.2.1 we have argued that General Problem Solvers or, in particular, General Vision Systems will never be available. Despite of that, our approach of boundary extraction is based on general compatibilities, the approach of object recognition is based on manifolds approximating general views, and our approach of robot control is based on a small set of generic modules. At first glance, the discussion in the introductory chapter seems to be in contradiction with the substantial chapters of the work. However, a detailed insight reveals that we do not strive for a General Vision System but strive for general development tools which mainly consist of statistical learning mechanisms. For example, the statistical evaluation of techniques for boundary extraction, the neural network learning of recognition functions, and the dynamic combination of deliberate and/or reactive vector fields serve as general development tools. Based on those, compatibilities, manifolds, and perception-action cycles can be learned for solving an exemplary robot vision task in the relevant environment. From an abstract point of view, the purpose of the development tools is to discover features which are more or less constant and features which change more or less continuous under the task-solving process. Additionally, covariances and courses of the two categories of features must be approximated which result in compatibilities and manifolds, respectively.

> Striving for general robot vision systems is hopeless and ridiculous, but striving for and applying general, learning-based development tools leads to really useful systems.

5.2 Rationale for the Contents of This Work

Work Does Not Present a Completely Working System

The primary purpose of this work is not to describe a working robot system which might has been developed for solving a specific Robot Vision task or a category of similar tasks.[1] Instead of that, the purpose has been to present a general methodology of designing and developing camera-equipped

---

[1] For persons interested in such systems I kindly recommend to visit my home page in the World Wide Web: http://www.ks.informatik.uni-kiel.de/~jpa/research.html .

robot systems. We focused on three essential issues, *i.e.* visual attention and boundary extraction in chapter 2, appearance-based recogniton of objects or situations in chapter 3, and perception-action cycles based on dynamic fields in chapter 4. The approaches were chosen exemplary, however, the main purpose has been to make clear the central role of learning. Therefore, the chapters are loosely coupled in order to facilitate the change of certain approaches by others which might be more appropriate for certain Robot Vision tasks.

Work Hardly Includes Benchmarking Tests

Generally, the work did not compare the presented approaches with other approaches (which is usually done with benchmarking tests). It was not our intention to develop new image processing techniques or robot steering strategies which might surpass a wide spectrum of already existing approaches. Instead of that, our intention has been to present a general methodology for automatically developing procedures which are acceptable for solving the underlying tasks. In the work, the only exception concerns Section 3.4 in which our favorite approach of object recognition has been compared with nearest neighbor classification. However, the reason was to clarify the important role of including a generalization bias in the process of learning.

Work Does Not Apply Specific, Vision-Related Models

The presented approaches for boundary extraction, object recognition, and robot control avoid the use of specific, vision-related models which might be derivable from specific, a priori knowledge. Rather, the intention has been to learn and to make use of general principles and thus to concentrate on minimalism principles. Based on that, it was interesting to determine the goodness of available solutions for certain Robot Vision tasks. Of course, there are numerous tasks to be performed in more or less customized environments which might be grounded on specific, vision-related models. The optical inspection of manufactured electronic devices is a typical example for this category of tasks. Nevertheless, in our opinion learning is necessary in any application to obtain a robust system. Therefore, our learning-based methodology should be revisited with the intention of including and making use of specific models, if available.

Work Is Written More or Less Informally

Formula are only introduced when they are necessary for understanding the approaches. Intentionally, we avoided an overload of the work with too much formalism. The universal use of formula would be meaningful if formal proofs for the correctness of techniques can be provided. However, the focus of this work is on applying learning mechanisms to Robot Vision. Unfortunately, it is an open issue of current research to prove the correctness of learning mechanisms or learned approximations.

5.3 Proposals for Future Research Topics

Infrastructure for the Experimentation Phase

The infrastructure for facilitating the experimentation phase must be improved. A technical equipment together with a comfortable user interface is needed for systematically changing environmental conditions in order to generate relevant variations of the input data and provide relevant training and test scenarios. Techniques are appreciated for propagating covariance from input data through a sequence of image processing algorithms and finally approximate the covariance of the output data [74]. Finally, we need a common platform for uniformly interchanging techniques which might be available from different research institutes throughout the world. For example, the Common Object Request Broker Architecture [121] could serve as a distributed platform to uniformly access a set of corner detectors, evaluate them, and finally choose the best one for the underlying task.

Fusion of Experimentation and Application Phase

Future work must deal with the problem of integrating the experimentation and the application phase thoroughly. Maybe, the designing phase and the application phase should be organized as a cycle, in which several iterations are executed on demand. In this case, the usability of certain task-solving strategies must be assessed by the system itself in order to execute a partial re-design automatically. However, scepticism is reasonable concerning the claim for an automatic design, because the evolutionary designing phase of the human brain did take billions of years to reach a deliberate level of solving high-level tasks.

Matter of Experimentations

In my opinion, it is more promising to extend the matter the experimental designing phase is dealing with. Most importantly, the system designer should make experiments with different learning approaches in the task-relevant environment in order to find out the most favorable ones. Based on this, the outcome of the experimental designing phase is not only a set of learned operators for object/situation recognition and a set of learned visual feedback mechanisms but the outcome could also be a learning approach itself, which is intended to learn operators for recognition and to adapt feedback mechanisms online.

Representational Integration of Perception and Action

In camera-equipped systems the robots can be used for two alternative purposes leading to a *robot-supported vision system (robot-for-vision tasks)* or to a *vision-supported robot system (vision-for-robot tasks)*. All techniques presented in our work are influenced by this confluence of perception and action. However, a more tight coupling is conveivable which might lead to an integrated representational space consisting of perceptive features and motor steering parameters. Consequently, the task-solving process would be determined and represented as a trajectory in the combined perception-action space.

Neural Network Learning for Robot Vision

Our work demonstrates exemplary the usefulness of Neural Networks for developing camera-equipped robot systems. Generally, the spectrum of possible applications of Neural Networks in the field of Robot Vision is far from being recognized. Future work on Robot Vision should increase the role to learning mechanisms both during the development and the application phase.

Appendix 1: Ellipsoidal Interpolation

Let $\Omega := \{X_1, \cdots, X_I\}$ be a set of vectors, which are taken from the m-dimensional vector space over the real numbers R. Based on the mean vector X^c of Ω we compute the matrix $\mathcal{M} := (X_1 - X^c, \cdots, X_I - X^c)$. The covariance matrix $\mathcal{C}$ of Ω is determined by the equation $\mathcal{C} = \frac{1}{I} \cdot \mathcal{M} \cdot \mathcal{M}^T$. For matrix $\mathcal{C}$ we obtain by principal component analysis the eigenvectors E_l and the corresponding eigenvalues λ_l, $l \in \{1, \cdots, I\}$. Let us assume to have the series of eigenvalues in decreasing order. Based on that the specific eigenvalue λ_I is equal to 0 and therefore eigenvector e_I can be cancelled (for an explanation, see Section 3.2). We define a canonical coordinate system (short, canonical frame) with the vector X^c as the origin and the eigenvectors $E_1, \cdots, E_{I-1}$ as the coordinate axes. The representation of vectors X_i, $i \in \{1, \cdots, I\}$, relative to the canonical frame is obtained by Karhunen-Loéve expansion according to equation (3.18). This yields a set of $I-1$-dimensional vectors $\hat{\Omega} := \{\hat{X}_1, \cdots, \hat{X}_I\}$. In the canonical frame we define an $(I-1)$-dimensional normal hyper-ellipsoid, i.e. principal axes are collinear with the axes of the canonical frame and the ellipsoid center is located at the origin of the frame. The half-lengths of this normal hyper-ellipsoid are taken as $\kappa_l := \sqrt{(I-1) \cdot \lambda_l}$, $l \in \{1, \cdots, I-1\}$.

Theorem 1 *All vectors X_i, $i \in \{1, \cdots, I\}$ are located on the specified hyper-ellipsoid.*

Proof *There are several $(I-1)$-dimensional hyper-ellipsoids which interpolate the set Ω of vectors, respectively. Principal component analysis determines the principal axes $E_1, \cdots, E_{I-1}$ of a specific hyper-ellipsoid which is subject to maximization of projected variances along candidate axes. Therefore, all corresponding vectors in $\hat{\Omega}$, which are represented in the canonical frame, are located on a normal hyper-ellipsoid with constant Mahalanobis distance h form the origin. With the given definition for the half-lengths we can show that h is equal to 1, which proves the theorem.*

Let the vectors in $\hat{\Omega}$ be defined as $\hat{X}_i := (\hat{x}_{i,1}, \cdots, \hat{x}_{i,I-1})^T$, $i \in \{1, \cdots, I\}$. From the vectors in $\hat{\Omega}$ the variance v_l along axis E_l, $l \in \{1, \cdots, I-1\}$ is given by $v_l := \frac{1}{I} \cdot (\hat{x}_{1,l}^2 + \cdots + \hat{x}_{I,l}^2)$. The variances v_l are equal to the eigenvalues λ_l. For each vector $\hat{X}_i$ we have the equation $\frac{\hat{x}_{i,1}^2}{\kappa_1^2} + \cdots + \frac{\hat{x}_{i,I-1}^2}{\kappa_{I-1}^2} = h$, because the vectors are located on a normal hyper-ellipsoid. Replacing κ_l^2 in the equation

J. Pauli: Learning-Based Robot Vision, LNCS 2048, pp. 263–264, 2001.
© Springer-Verlag Berlin Heidelberg 2001

by the expression $\frac{I-1}{I} \cdot \left(\hat{x}_{1,l}^2 + \cdots + \hat{x}_{I,l}^2 \right)$ yields the following equation $\frac{I}{I-1} \cdot \left(\frac{\hat{x}_{i,1}^2}{\hat{x}_{1,1}^2 + \cdots + \hat{x}_{I,1}^2} + \cdots + \frac{\hat{x}_{i,I-1}^2}{\hat{x}_{1,I-1}^2 + \cdots + \hat{x}_{I,I-1}^2} \right) = h$. Summing up all these equations for $i \in \{1, \cdots, I\}$ yields the equation $\frac{I}{I-1} \cdot (I-1) = I \cdot h$, which results in $h = 1$.

q.e.d.

Appendix 2: Further Behavioral Modules

In the following we present the behavioral modules MB_4, MB_5, and MB_6. They are more sophisticated compared to MB_1, MB_2, and MB_3, which have been introduced in Subsection 4.2.2.

Generic Module MB_4 for Assembled Behavior

The module is similar to the generic module MB_2 in that the state of the effector must be changed continually under the constraint of keeping desired image measurements. However, instead of combining two goal-oriented cycles, one replaces the outer cycle by an *exploration strategy*, *i.e.* the module is intended to solve an exploration sub-task. Only one type of image measurements is needed, and the robot effector keeps on changing its variable state while trying to keep desired measurements (*e.g.* wandering along a wall). For this purpose, the effector continually changes its variable state until a desired measurement in the images is obtained (*e.g.* proximity to the wall), then changes the state of the effector for doing an exploration step (*e.g.* hypothetical step along the wall), and once again the effector is controlled to reach desired measurements (*e.g.* coming back to the wall but at a displaced position). In order to avoid to frequently come back into the same state vector, the state vector belonging to a desired measurement is memorized, and based on this one specifies constraints for the exploration step, *e.g.* extrapolating from approximated history. An elementary behavior implemented by behavioral module MB_1 is included, but no plan in form of a deliberate field is used. Instead, information about the relevant course of effector states is produced and is represented as a sequence of equilibrium points in a deliberate field. This is an example of bottom-up flow of information (*e.g.* course of the wall).

J. Pauli: Learning-Based Robot Vision, LNCS 2048, pp. 265-267, 2001.
© Springer-Verlag Berlin Heidelberg 2001

Module MB_4

---

1. Determine relevant type of variable state vectors $\wr S^v \wr$ and accompanying type of measurements $\wr Q \wr$.
   Initialization of a deliberate field for $\wr S^v \wr$.
2. Behavioral module MB_1:
   Configure with ($\wr S^v \wr$, $\wr Q \wr$), execution, and return ($S^v(t)$).
3. Construct an equilibrium in the deliberate field based on current state vector $S^v(t)$.
4. Approximate recent history of equilibrium points and virtually extrapolate the course of state vectors by a small increment vector.
5. By instruction change the variable state vector into the virtually extrapolated state, and increment time parameter $t := t + 1$.
6. If (the_new_state_vector_has_reached_a_certain_relation_ with_the_whole_history_of_equilibrium_points) then go to 8.
7. Go to 2.
8. Memorize final deliberate field $\wr S^v \wr$, and stop.

Generic Module MB_5 for Assembled Behavior

The module is responsible for an assembled behavior which integrates *deliberate plan execution* and *reactive fine-tuning*. The rough behavior for the effector is based on a short-term plan which is represented by a deliberate field. For example, the deliberate field can be composed of one attractor and several repellors. In an outer cycle a small change of the effector state is executed according to the plan. Then, in the inner cycle the behavioral module MB_1 must fine-tune the effector state such that desired image measurements are reached. Next, again a deliberate step is executed according to the plan, and so on. The procedure is repeated until an equilibrium is reached in the deliberate field.

Module MB_5

---

1. Determine relevant type of variable state vectors $\wr S^v \wr$ and accompanying type of measurements $\wr Q \wr$.
   Take deliberate field of $\wr S^v \wr$ into account.
2. Behavioral module MB_1:
   Configure with ($\wr S^v \wr$, $\wr Q \wr$), execution, and return ($S^v(t)$).
3. Determine current state vector $S^v(t)$ of the relevant effector.
4. Determine control vector according to equation
   $C(t) := VF_O[\{S^v_A\}, \{S^v_{Rj}\}](S^v(t))$.
5. If ($\| C(t) \| \leq \eta_1$) then go to 8.

Module MB_5, continued

---

6. Change variable state vector according to equation
   $S^v(t+1) := f^{ts}(C(t), S^v(t))$, and increment time parameter
   $t := t + 1$.
7. Go to 2.
8. Stop.

Generic Module MB_6 for Assembled Behavior

The module is responsible for an assembled behavior which also combines *deliberate plan execution* and *visual feedback control*. However, in distinction to the previous module MB_5 now *two effectors* are involved. The first effector is working according to the plan, and the second effector is controlled by visual feedback. For example, a vehicle is supposed to move to a certain position, which is done by a plan, and simultaneously an agile head-camera is supposed to track a by-passing object, which is done by visual feedback control. The planned vehicle movement may be useful if the goal position is far away and/or does not contain a landmark, and consequently the goal position can not be determined from images. For cases like those, the goal position must be determined based on task specification.

Module MB_6

---

1. Determine relevant type of variable state vectors $\wr S_1^v \wr$.
   Take deliberate field of $\wr S_1^v \wr$ into account.
   Determine relevant type of variable state vectors $\wr S_2^v \wr$ and accompanying type of measurements $\wr Q_2^v \wr$.
   Initialization of a deliberate field for $\wr S_2^v \wr$.
2. Behavioral module MB_1:
   Configure with ($\wr S_2^v \wr$, $\wr Q_2^v \wr$), execution, and return ($S_2^v(t)$).
3. Construct an equilibrium in the deliberate field $\wr S_2^v \wr$ based on current state vector $S_2^v(t)$.
4. Determine current state vector $S_1^v(t)$ of the relevant effector.
5. Determine control vector according to equation
   $C_1^v(t) := VF_O[\{S_{1\,A}^v\}, \{S_{1\,Rj}^v\}](S_1^v(t))$.
6. If ($\parallel C_1^v(t) \parallel \leq \eta_1$) then go to 9.
7. Change variable state vector according to equation
   $S_1^v(t+1) := f^{ts}(C_1^v(t), S_1^v(t))$, and increment time parameter
   $t := t + 1$.
8. Go to 2.
9. Memorize final deliberate field $\wr S_2^v \wr$, and stop.

Symbols

Symbols for Scalars

$i, j, \cdots$	Indices	*
$I, J, \cdots$	Upper bounds of indices	*
I_w, I_h, I_d	Width, height, diagonal of an image	32
x_i, y_i	Image and scene coordinates	32
J_w, J_h	Width and height of LPT image	137
v_1, v_2	Coordinates of LPT image	137
r	Distance of a line from image center	33
ϕ	Orientation of a line	33
b	Effective focal length	47
$\delta_i, \eta_i, \zeta_i, \psi$	Threshold parameters for various purposes	*
λ_i	Weighting factors for compatibilities	*
s	Factor for stepwise effector movement	221
ξ	Normalizing factor of projection equation	245
μ	Regularization parameter	112
ζ	Threshold in PAC methodology	110
P^r	Probability in PAC methodology	111
v_A, v_B	Parameters of virtual force vector field	185
φ	Phase angle in a complex plane	49
$\alpha_i, \beta_i, \gamma_i, \phi_i$	Angles for lines and junctions	*
l_i^s	Lengths of line segments	55
ρ	Radius of polar coordinates	137
θ	Angle of polar coordinates	137
ρ^{min}	Radius of a foveal image component	137
ρ^{max}	Radius of a peripheral image component	137
κ_i	Half-lengths of axes of hyper-ellipsoid	150
σ_i, τ	Parameters of a GBF	112
w_i	Scalar weights	*
λ_i	Eigenvalues of covariance matrix for PCA	118
c_r	Radial center frequency of Gabor function	85

Symbols for Sets

$\mathcal{R}, \mathcal{R}^m$	Set of real numbers, m-dim. vector space	112
$\mathcal{P}, \mathcal{P}_S$	Set or subset of image coordinate tuples	32

Symbols for Sequences

Symbols for Images

Symbols for Vector Fields

Symbols for Matrices

Symbols for Vectors

Y	Output vector	110
B	Parameter vector	109
X^c	Mean vector from an ensemble	*
$\mathcal{U}$	Center frequencies of Gabor function	36
$q, p, p_a, \cdots$	Image coordinate tuples	39
P_i	Scene coordinate tuples	*
$Q(t)$	Current measurement vector	188
Q^*	Desired measurement vector	188
S^c	Fixed (constant) state vector	183
$S^v(t)$	Variable (current) state vector	183
$C(t)$	Control vector	184
S_A^v	Variable state vector defining an attractor	184
S_R^v	Variable state vector defining a repellor	185
Δ_i^1, Δ_i^2	Difference vectors	150
X_i^s, X_i^v	Seed vector, validation vector	155
E_i	Eigenvectors of covariance matrix for PCA	118
W	Weight vector for GBF network	113

Symbols for Functions

$f, \tilde{f}$	Explicit functions	112
f^{im}, f^{Gi}	Implicit functions	109
f^L	Function of polar line representation	33
f^{Gs}	Gauss function	112
f_i^{mg}	Functions for modifying Gaussians	150
f_i^{Gm}	Specific, hyper-ellipsoidal Gaussian	150
$f_{\mathcal{D},\mathcal{U}}^{Gb}$	Gabor function	36
f^{ph}	Computing mean local phase of line segment	53
f_k^{rc}	Recognition functions	157
f^{ts}	Transition function for variable state vector	184
f^{v1}, f^{v2}	Computing coordinates of LPT image	138
f^z	Determine LPT related number of image pixels	139
f^{ms}	Measurement function applied to an image	187
f^{ct}	Control function generating a control vector	188
f^{li}, f^{nl}	Projection functions	245
F, F^{sm}	Functionals	112

Symbols for Deviation/Simularity Measures

D_{OO}	Distance between angles modulo 180	39
D_{LE}	Orientation-deviation of a line segment	39
D_{PC}	Junction-deviation of a line pencil	42
D_{JP}	Euclidean distance between positions	42
D_{JO}	Deviation between sequences of angles	42
D_{PR}	Similarity between local phases	53

D_{RC}	Rectangle-deviation of a quadrangle	56
D_{PA}	Parallelogram-deviation of a quadrangle	56
D_{SQ}	Square-deviation of a quadrangle	56
D_{RH}	Rhombus-deviation of a quadrangle	56
D_{TR}	Trapezoid-deviation of a quadrangle	56
D_{LE_QD}	Orientation-deviation of a quadrangle	55
D_{CC_QD}	Junction-deviation related to a quadrangle	55
D_{PR_QD}	Phase-similarity of a quadrangle	55
D_{SP_QD}	Generic measure for quadrangle deviation	56
D_{LE_PG}	Orientation-deviation of a polygon	61
D_{CC_PG}	Junction-deviation of a polygon	61
D_{PR_QD}	Phase-similarity related to a polygon	68
D_{SP_PG}	Generic measure for polygon deviation	68
D_{RS}, d_{rs}	Deviation from reflection-symmetry	66
D_{TS}, d_{ts}	Deviation from translation-symmetry	66
D_{RA}, d_{ra}	Deviation from a right-angled polygon	67

Symbols for Other Measures

A_{SP_QD}	Saliency measure of a specific quadrangle	56
A_{SP_PG}	Saliency measure of a specific polygon	68
V_{SL}	Normalized length variance of line segments	56

Symbols for Methods

PE_1	Generic procedure for quadrangle extraction	57
PE_2	Generic procedure for polygon extraction	68
PE_3	Generic procedure for polyhedra extraction	78
PE_4	Generic procedure for parallelepiped extraction	81
CF_{1NN}	Object recognition with 1-nearest neighbor	162
CF_{ELL}	Object recognition with ellipsoid approximation	162
CF_{EGN}	Object recognition with GBF/ellipsoid network	162
MI_i	Instructional modules	198
MB_i	Behavioral modules	200
MM_i	Monitoring modules	203
MT_i	Task-specific modules	212

Other Symbols

M-junction	Junction with M converging lines	42
$\wr S^v \wr$	Type of variable state vector	199
$\wr Q \wr$	Type of measurement vector	200
CA_i	Camera designations	207
$\hat{i}$	Imaginary unit	36

Index

References

1. Y. Aloimonos, editor: *Active Perception.* Lawrence Erlbaum Associates Publishers, Hillsdale, New Jersey, 1993.

2. Y. Aloimonos: Active vision revisited. In Y. Aloimonos, editor, *Active Perception,* pages 1–18, Lawrence Erlbaum Associates Publishers, Hillsdale, New Jersey, 1993.

3. A. Amir and M. Lindenbaum: A generic grouping algorithm and its quantitative analysis. In *IEEE Transactions on Pattern Analysis and Machine Intelligence,* volume 20, pages 168–185, 1998.

4. Anonymous, editor: *IEEE Workshop on Qualitative Vision.* IEEE Computer Society Press, Los Alamitos, California, 1993.

5. Anonymous, editor: *AAAI Fall Symposium on Machine Learning in Computer Vision,* AAAI Press, Menlo Park, California, 1993.

6. M. Anthony: *Probabilistic Analysis of Learning in Artificial Neural Networks – The PAC Model and its Variants.* Technical report NC-TR-94-3, The London School of Economics and Political Science, Department of Mathematics, England, 1994.

7. R. Arkin: *Behavior-based Robotics.* The MIT Press, Cambridge, Massachusetts, 1998.

8. D. Arnold, K. Sturtz, V. Velten, and N. Nandhakumar: Dominant-subspace invariants. In *IEEE Transactions on Pattern Analysis and Machine Intelligence,* volume 22, pages 649–662, 2000.

9. R. Bajcsy and M. Campos: Active and exploratory perception. In *Computer Vision and Image Understanding,* volume 56, pages 31–40, 1992.

10. D. Ballard: Generalizing the Hough transform to detect arbitrary shapes. In *Pattern Recognition,* volume 13, pages 111–122, 1981.

11. D. Ballard and C. Brown: *Computer Vision.* Prentice-Hall, Englewood Cliffs, New Jersey, 1982.

12. D. Ballard and C. Brown: Principles of animate vision. In *Computer Vision and Image Understanding,* volume 56, pages 3–21, 1992.

13. S. Baluja: Evolution of an artifical neural network based autonomous land vehicle controller. In *IEEE Transactions on Systems, Man, and Cybernetics,* volume 26, pages 450–463, 1996.

14. A. Barto, S. Bradtke, and S. Singh: Learning to act using real-time dynamic programming. In *Artificial Intelligence,* volume 72, pages 81–138, 1995.

15. E. Baum and D. Haussler: What net size gives valid generalization ? In J. Shavlik and T. Dietterich, editors, *Readings in Machine Learning*, pages 258–262, Morgan Kaufmann Publishers, San Francisco, California, 1990.

16. S. Becker: Implicit learning in 3D object recognition – The importance of temporal context. In *Neural Computation*, volume 11, pages 347–374, 1999.

17. P. Belhumeur and D. Kriegman: What is the set of images of an object under all possible lighting conditions ? In *IEEE Conference on Computer Vision and Pattern Recognition*, pages 270–277, IEEE Computer Society Press, Los Alamitos, California, 1996.

18. J. Benediktsson, J. Sveinsson, O. Ersoy, and P. Swain: Parallel consensual neural networks. In *IEEE Transactions on Neural Networks*, volume 8, pages 54–64, 1997.

19. B. Bhanu and T. Poggio, editors: *IEEE Transactions on Pattern Analysis and Machine Intelligence*. Special journal issue on Learning in Computer Vision, volume 16, pages 865–960, IEEE Computer Society Press, Los Alamitos, California, 1994.

20. T. Binford and T. Levitt: Quasi-invariants – Theory and exploitation. In *Image Understanding Workshop*, pages 819–829, Morgan Kaufmann Publishers, San Francisco, California, 1993.

21. Ch. Bishop: *Neural Networks for Pattern Recognition*. Clarendon Press, London, England, 1995.

22. W. Blase: *RBF basierte Neuronale Netze zum automatischen Erwerb hindernisfreier Manipulator-Trajektorien, basierend auf monokularen Bildfolgen.* Master thesis (in german), Christian-Albrechts-Universität zu Kiel, Institut für Informatik und Praktische Mathematik, Germany, 1998.

23. M. Bolduc and M. Levine: A review of biologically motivated space-variant data reduction models for Robot Vision. In *Computer Vision and Image Understanding*, volume 69, pages 170–184, 1998.

24. M. Borga: *Learning Multidimensional Signal Processing*. PhD thesis, Linköping University, Department of Electrical Engineering, Sweden, 1998.

25. R. Brooks: A robust layered control system for a mobile robot. In *IEEE Journal of Robotics and Automation*, volume 2, pages 14–23, 1986.

26. R. Brooks: Intelligence without representation. In *Artificial Intelligence*, volume 47, pages 139–159, 1991.

27. R. Brooks: Intelligence without reason. In *Proceedings of the Twelfth International Joint Conference on Artificial Intelligence*, pages 569–595, Morgan Kaufmann Publishers, San Francisco, California, 1991.

28. R. Brooks: New approaches to robotics. In *Science*, volume 253, pages 1227–1232, 1993.

29. J. Bruske and G. Sommer: Dynamic cell structure learns perfectly topology preserving map. In *Neural Computation*, volume 7, pages 845–865, 1995.

30. J. Bruske and G. Sommer: Intrinsic dimensionality estimation with optimally topology preserving maps. In *IEEE Transactions on Pattern Analysis and Machine Intelligence*, volume 20, pages 572–575, 1998.

31. C. Burges: *A Tutorial on Support Vector Machines for Pattern Recognition.* URL http://www.kernel-machines.org/papers/Burges98.ps.gz, 1998.

32. C. Burges: Geometry and invariance in kernel based methods. In B. Schölkopf, C. Burges, and A. Smola, editors, *Advances in Kernel Methods*, pages 89–116, The MIT Press, Cambridge, Massachusetts, 1998.

33. A. Castano and S. Hutchinson: A probabilistic approach to perceptual grouping. In *Computer Vision and Image Understanding*, volume 64, pages 399–419, 1996.

34. R. Chatila: Deliberation and reactivity in autonomous mobile robots. In *Robotics and Autonomous Systems*, volume 16, pages 197–211, 1995.

35. F. Chaumette, S. Boukir, P. Bouthemy, and D. Juvin: Structure from controlled motion. In *IEEE Transactions on Pattern Analysis and Machine Intelligence*, volume 18, pages 492–504, 1996.

36. K. Cho and P. Meer: Image segmentation from consensus information. In *Computer Vision and Image Understanding*, volume 68, pages 72–89, 1997.

37. H. Christensen and W. Förstner, editors: *Machine Vision and Applications.* Special journal issue on Performance Characteristics of Vision Algorithms, volume 9, pages 215–340, Springer Verlag, Berlin, 1997.

38. D. Cohn, L. Atlas, and R. Ladner: Improving generalization with active learning. In *Machine Learning*, volume 15, pages 201–221, 1994.

39. M. Colombetti, M. Dorigo, and G. Borghi: Behavior analysis and training – A methodology for behavior engineering. In *IEEE Transactions on Systems, Man, and Cybernetics*, volume 26, pages 365–380, 1996.

40. C. Colombo, M. Rucci, and P. Dario: Attentive behavior in an anthropomorphic Robot Vision system. In *Robotics and Autonomous Systems*, volume 12, pages 121–131, 1994.

41. P. Corke: Visual control of robot manipulators – A review. In K. Hashimoto, editor, *Visual Servoing*, World Scientific Publishing, Singapore, pages 1–31, 1993.

42. J. Craig: *Introduction to Robotics.* Addison-Wesley Publishing Company, Reading, Massachusetts, 1989.

43. D. Crevier and R. Lepage: Knowledge-based image understanding systems – A survey. In *Computer Vision and Image Understanding*, volume 67, pages 161–185, 1997.

44. M. Cutkosky: On grasp choice, grasp models, and the design of hands for manufacturing tasks. In *IEEE Transactions on Robotics and Automation*, volume 5, pages 269–279, 1989.

45. A. Cypher, editor: *Watch What I Do – Programming by Demonstration.* The MIT Press, Cambridge, Massachusetts, 1993.

46. P. Dayan, G. Hinton, R. Neal, and R. Zemel: The Helmholtz machine. In *Neural Computation*, volume 7, pages 889–904, 1995.

47. S. Dickinson, H. Christensen, J. Tsotsos, and G. Olofsson: Active object recognition integrating attention and viewpoint control. In *Computer Vision and Image Understanding*, volume 67, pages 239–260, 1997.

48. J.-Y. Donnart and J.-A. Meyer: Learning reactive and planning rules in a motivationally autonomous animat. In *IEEE Transactions on Systems, Man, and Cybernetics*, volume 26, pages 381–395, 1996.

49. M. Dorigo, editor: *IEEE Transactions on Systems, Man, and Cybernetics*. Special journal issue on Learning Autonomous Robots, volume 26, pages 361–505, The IEEE Inc., New York, 1996.

50. C. Engels and G. Schöner: Dynamic fields endow behavior-based robots with representations. In *Robotics and Autonomous Systems*, volume 14, pages 55–77, 1995.

51. M. Erdmann: Understanding action and sensing by designing action-based sensors. In *The International Journal of Robotics Research*, volume 14, pages 483–509, 1995.

52. O. Faugeras: *Three-Dimensional Computer Vision*. The MIT Press, Cambridge, Massachusetts, 1993.

53. O. Faugeras: Stratification of three-dimensional vision – Projective, affine and metric representations. In *Journal of the Optical Society of America*, volume 12, pages 465–484, 1995.

54. J. Feddema, C. Lee, and O. Mitchell: Model-based visual feedback control for a hand-eye coordinated robotic system. In *Computer*, pages 21–31, 1992.

55. G. Finlayson, B. Funt, and K. Barnard: Color constancy under varying illumination. In *International Conference on Computer Vision*, pages 720–725, IEEE Computer Society, Los Alamitos, California, 1995.

56. S. Floyd and M. Warmuth: Sample compression, learnability, Vapnik–Chervonenkis dimension. In *Machine Learning*, volume 21, pages 269–304, 1995.

57. G. Foresti, V. Murino, C. Regazzoni, and G. Vernazza: Grouping of rectilinear segments by the labeled Hough transform. In *Computer Vision and Image Understanding*, volume 58, pages 22–42, 1994.

58. H. Friedrich, S. Münch, and R. Dillmann: Robot programming by demonstration – Supporting the induction by human interaction. In *Machine Learning*, volume 23, pages 163–189, 1996.

59. B. Fritzke: Growing cell structures – A self-organizing network for unsupervised and supervised learning. In *Neural Networks*, volume 7, pages 1441–1460, 1994.

60. K. Fukunaga: *Introduction to Statistical Pattern Recognition*. Academic Press, San Diego, California, 1990.

61. E. Gamma, R. Helm, R. Johnson, and J. Vlissides: *Design Patterns*. Addison-Wesley Publishing Company, Reading, Massachusetts, 1995.

62. S. Geman, E. Bienenstock, and R. Doursat: Neural networks and the bias/variance dilemma. In *Neural Computation*, volume 4, pages 1–58, 1992.

63. F. Girosi, M. Jones, and T. Poggio: Regularization theory and neural network architectures. In *Neural Computation*, volume 7, pages 219–269, 1995.

64. F. Girosi: An equivalence between sparse approximation and support vector machines. In *Neural Computation*, volume 10, pages 1455–1480, 1998.

65. L. Goldfarb, S. Deshpande, and V. Bhavsar: *Inductive Theory of Vision.* Technical report 96-108, University of New Brunswick, Faculty of Computer Science, Canada, 1996.

66. G. Granlund and H. Knutsson: *Signal Processing for Computer Vision.* Kluwer Academic Publishers, Dordrecht, The Netherlands, 1995.

67. R. Greiner and R. Isukapalli: Learning to select useful landmarks. In *IEEE Transactions on Systems, Man, and Cybernetics*, volume 26, pages 437–449, 1996.

68. P. Gros, O. Bournez, and E. Boyer: Using local planar geometric invariants to match and model images of line segments. In *Computer Vision and Image Understanding*, volume 69, pages 135–155, 1998.

69. S. Gutjahr and J. Feist: Elliptical basis function networks for classification tasks. In *International Conference on Artificial Neural Networks*, Lecture Notes in Computer Science, volume 1327, pages 373–378, Springer Verlag, Berlin, 1997.

70. G. Hager, W. Chang, and A. Morse: Robot hand-eye coordination based on stereo vision. In *IEEE Control Systems*, pages 30–39, 1995.

71. P. Hall, D. Marshall, and R. Martin: Merging and splitting Eigenspace models. In *IEEE Transactions on Pattern Analysis and Machine Intelligence*, volume 22, pages 1042–1049, 2000.

72. R. Haralick and L. Shapiro: *Computer and Robot Vision.* Volume I, Addison-Wesley Publishing Company, Reading, Massachusetts, 1993.

73. R. Haralick and L. Shapiro: *Computer and Robot Vision.* Volume II, Addison-Wesley Publishing Company, Reading, Massachusetts, 1993.

74. R. Haralick: Propagating covariance in Computer Vision. In R. Klette, S. Stiehl, M. Viergever, and K. Vincken, editors, *Evaluation and Validation of Computer Vision Algorithms*, pages 95–114, Kluwer Academic Publishers, Dordrecht, The Netherlands, 2000.

75. S. Harnad: The symbol grounding problem. In *Physica D*, volume 42, pages 335–346, 1990.

76. K. Hashimoto, editor: *Visual Servoing.* World Scientific Publishing, Singapore, 1993.

77. P. Havaldar, G. Medioni, and F. Stein: Perceptual grouping for generic recognition. In *International Journal of Computer Vision*, volume 20, pages 59–80, 1996.

78. R. Horaud and F. Dornaika: Hand-Eye calibration. In *International Journal of Robotics Research*, volume 14, pages 195–210, 1995.

79. R. Horaud and F. Chaumette, editors: *International Journal of Computer Vision.* Special journal issue on Image-based Robot Servoing, volume 37, pages 5–118, Kluwer Academic Publishers, Dordrecht, The Netherlands, 2000.

80. B. Horn and B. Schunck: Determining optical flow. In *Artificial Intelligence*, volume 17, pages 185–203, 1981.

81. K. Hornik, M. Stinchcombe, and H. White: Multilayer feedforward networks are universal approximators. In *Neural Networks*, volume 2, pages 359–366, 1989.

82. S. Hutchinson, G. Hager, and P. Corke: A tutorial on visual servo control. In *IEEE Transactions on Robotics and Automation*, volume 12, pages 651–670, 1996.

83. K. Ikeuchi and T. Kanade: Automatic generation of object recognition programs. In *Proceedings of the IEEE*, volume 76, pages 1016–1035, 1988.

84. K. Ikeuchi and T. Suehiro: Toward an assembly plan from observation – Part I – Task recognition with polyhedral objects. In *IEEE Transactions on Robotics and Automation*, volume 10, pages 368–385, 1994.

85. B. Jähne: *Practical Handbook on Image Processing for Scientific Applications*. CRC Press, Boca Raton, Florida, 1997.

86. R. Jain and T. Binford: Ignorance, myopia, and naiveté in Computer Vision systems. In *Computer Vision and Image Understanding*, volume 53, pages 112–117, 1991.

87. B. Julesz: Early vision is bottom-up, except for focal attention. *Cold Spring Harbor Symposia on Quantitative Biology*, volume LV, pages 973–978, Cold Spring Harbor Laboratory Press, Cold Spring Harbor, New York, 1990.

88. L. Kaelbling, editor: *Machine Learning*. Special journal issue on Reinforcement Learning, volume 22, pages 5–290, Kluwer Academic Publishers, Dordrecht, The Netherlands, 1996.

89. N. Kambhatla and T. Leen: Fast nonlinear dimension reduction. In J. Cowan, G. Tesauro, and J. Alspector, editors, *Advances in Neural Information Processing Systems*, volume 6, pages 152–159, Morgan Kaufmann Publishers, San Francisco, California, 1993.

90. R. Klette, K. Schlüns, and A. Koschan: *Computer Vision – Three-Dimensional Data from Images*. Springer Verlag, Berlin, 1998.

91. E. Kruse and F. Wahl: Camera-based monitoring system for mobile robot guidance. In *IEEE International Conference on Intelligent Robots and Systems*, The IEEE Inc., New York, 1998.

92. S. Kulkarni, G. Lugosi, and S. Venkatesh: Learning pattern classification – A survey. In *IEEE Transactions on Information Theory*, volume 44, pages 2178–2206, 1998.

93. Y. Kuniyoshi, M. Inaba, and H. Inoue: Learning by watching – Extracting reusable task knowledge from visual observation of human performance. In *IEEE Transactions on Robotics and Automation*, volume 10, pages 799–822, 1994.

94. S. Kunze: *Ein Hand-Auge-System zur visuell basierten Lokalisierung und Identifikation von Objekten*. Master thesis (in german), Christian-Albrechts-Universität zu Kiel, Institut für Informatik und Praktische Mathematik, Germany, 1999.

95. S. Kurihara, S. Aoyagi, R. Onai, and T. Sugawara: Adaptive selection of reactive/deliberate planning for a dynamic environment. In *Robotics and Autonomous Systems*, volume 24, pages 183–195, 1998.

96. K. Kutulakos and C. Dyer: Global surface reconstruction by purposive control of observer motion. In *Artificial Intelligence*, volume 78, pages 147–177, 1995.

97. C. Lam, S. Venkatesh, and G. West: Hypothesis verification using parametric models and active vision strategies. In *Computer Vision and Image Understanding*, volume 68, pages 209–236, 1997.

98. M. Landy, L. Maloney, and M. Pavel, editors: *Exploratory Vision – The Active Eye*. Springer Verlag, Berlin, 1995.

99. J.-C. Latombe: *Robot Motion Planning*. Kluwer Academic Publishers, Dordrecht, The Netherlands, 1991.

100. V. Leavers: Survey – Which Hough transform ? In *Computer Vision and Image Understanding*, volume 58, pages 250–264, 1993.

101. S. Li: Parameter estimation for optimal object recognition – Theory and application. In *International Journal of Computer Vision*, volume 21, pages 207–222, 1997.

102. C.-E. Liedtke and A. Blömer: Architecture of the knowledge-based configuration system for image analysis – CONNY. In *IAPR International Conference on Pattern Recognition*, pages 375–378, IEEE Computer Society Press, Los Alamitos, California, 1997.

103. D. Lowe: *Perceptual Organization and Visual Recognition*. Kluwer Academic Publishers, Dordrecht, The Netherlands, 1985.

104. É. Marchand and F. Chaumette: Controlled camera motions for scene reconstruction and exploration. In *IEEE Conference on Computer Vision and Pattern Recognition*, pages 169–176, IEEE Computer Society Press, Los Alamitos, California, 1996.

105. D. Marr: *Vision – A Computational Investigation into the Human Representation and Processing of Visual Information*. Freeman and Company, New York, 1982.

106. T. Martinetz and K. Schulten: Topology representing networks. In *Neural Networks*, volume 7, pages 505–522, 1994.

107. J. Matas, J. Burianek, and J. Kittler: Object recognition using the invariant pixel-set signature. In M. Mirmehdi and B. Thomas, editors, *British Machine Vision Conference*, volume 2, pages 606–615, ILES Central Press, Bristol, England, 2000.

108. D. Metaxas and D. Terzopoulos: *Computer Vision and Image Understanding*. Special journal issue on Physics-Based Modeling and Reasoning in Computer Vision, volume 65, pages 111–360, Academic Press, San Diego, California, 1997.

109. J. Müller: A cooperation model for autonomous agents. In J. Müller, M. Wooldridge, and N. Jennings, editors, *Intelligent Agents III*, Lecture Notes in Artificial Intelligence, volume 1193, pages 245–260, Springer Verlag, Berlin, 1997.

110. J. Mundy and A. Zisserman, editors: *Geometric Invariance in Computer Vision*. The MIT Press, Cambridge, Massachusetts, 1992.

111. J. Mundy and A. Zisserman: Introduction – Towards a new framework for vision. In J. Mundy and A. Zisserman, editors, *Geometric Invariance in Computer Vision*, pages 1–39, The MIT Press, Cambridge, Massachusetts, 1992.

112. H. Murase and S. Nayar: Visual learning and recognition of 3D objects from appearance. In *International Journal of Computer Vision*, volume 14, pages 5–24, 1995.

113. R. Murphy, D. Hershberger, and G. Blauvelt: Learning landmark triples by experimentation. In *Robotics and Autonomous Systems*, volume 22, pages 377–392, 1997.

114. R. Murphy, K. Hughes, A. Marzilli, and E. Noll: Integrating explicit path planning with reactive control of mobile robots using Trulla. In *Robotics and Autonomous Systems*, volume 27, pages 225–245, 1999.

115. F. Mussa-Ivaldi: From basis functions to basis fields – Vector field approximation from sparse data. In *Biological Cybernetics*, volume 67, pages 479–489, 1992.

116. S. Nayar and T. Poggio, editors: *Early Visual Learning*. Oxford University Press, Oxford, England, 1996.

117. S. Negahdaripour and A. Jain: *Final Report of the NSF Workshop on the Challenges in Computer Vision Research*. University of Miami, Florida, 1991.

118. A. Newell and H. Simon: GPS – A program that simulates human thought. In E. Feigenbaum and J. Feldman, editors, *Computers and Thought*, pages 279–293, McGraw-Hill, New York, 1963.

119. N. Nilsson: *Principles of Artificial Intelligence*. Tioga Publishing, Palo Alto, California, 1980.

120. E. Oja: *Subspace Methods of Pattern Recognition*. Research Studies Press, Hertfordshire, England, 1983.

121. R. Orfali and D. Harkey: *Client/Server Programming with JAVA and CORBA*. Wiley Computer Publishing, New York, 1998.

122. M. Orr: *Introduction to Radial Basis Function Networks*. University of Edinburgh, Institute for Adaptive and Neural Computation, URL http://www.anc.ed.ac.uk/~mjo/papers/intro.ps.gz, 1996.

123. P. Palmer, J. Kittler, and M. Petrou: An optimizing line finder using a Hough transform algorithm. In *Computer Vision and Image Understanding*, volume 67, pages 1–23, 1997.

124. P. Papanikolopoulos and P. Khosla: Robotic visual servoing around a static target – An example of controlled active vision. In *Proceedings of the American Control Conference*, pages 1489–1494, 1992.

125. T. Papathomas and B. Julesz: Lie differential operators in animal and machine vision. In J. Simon, editor, *From Pixels to Features*, pages 115–126, Elsevier Science Publishers, Amsterdam, The Netherlands, 1989.

126. J. Pauli: Recognizing 2D image structures by automatically adjusting matching parameters. In *German Workshop on Artificial Intelligence (GWAI)*, Informatik Fachberichte, volume 251, pages 292–296, Springer Verlag, Berlin, 1990.

127. J. Pauli, M. Benkwitz and G. Sommer: RBF networks for object recognition. In B. Krieg-Brueckner and Ch. Herwig, editors, *Workshop Kognitive Robotik*, Zentrum für Kognitive Systeme, Universität Bremen, ZKW-Bericht 3/95, 1995.

128. U. Pietruschka and M. Kinder: Ellipsoidal basis functions for higher-dimensional approximation problems. In *International Conference on Artificial Neural Networks*, volume 2, pages 81–85, EC2 et Cie, Paris, France, 1995.

129. P. Pirjanian: *Multiple Objective Action Selections and Behavior Fusion using Voting*. PhD thesis, Aalborg University, Department of Medical Informatics and Image Analysis, Denmark, 1998.

130. T. Poggio and F. Girosi: Networks for approximation and learning. In *Proceedings of the IEEE*, volume 78, pages 1481–1497, 1990.

131. T. Poggio and F. Girosi: A sparse representation for function approximation. In *Neural Computation*, volume 10, pages 1445–1454, 1998.

132. M. Pontil and A. Verri: Support vector machines for 3D object recognition. In *IEEE Transactions on Pattern Analysis and Machine Intelligence*, volume 20, pages 637–646, 1998.

133. M. Prakash and M. Murty: Extended sub-space methods of pattern recognition. In *Pattern Recognition Letters*, volume 17, pages 1131–1139, 1996.

134. W. Press, S. Teukolsky, and W. Vetterling: *Numerical Recipes in C*. Cambridge University Press, Cambridge, Massachusetts, 1992.

135. J. Princen, J. Illingworth, and J. Kittler: A hierarchical approach to line extraction based on the Hough transform. In *Computer Vision, Graphics, and Image Processing*, volume 52, pages 57–77, 1990.

136. R. Rao and D. Ballard: Object indexing using an iconic sparse distributed memory. In *International Conference on Computer Vision*, pages 24–31, IEEE Computer Society, Los Alamitos, California, 1995.

137. K. Ray and D. Majumder: Application of Hopfield neural networks and canonical perspectives to recognize and locate partially occluded 3D objects. In *Pattern Recognition Letters*, volume 15, pages 815–824, 1994.

138. O. Rioul and M. Vetterli: Wavelets and signal processing. In *IEEE Signal Processing Magazine*, pages 14–38, 1991.

139. J. Rissanen: Universal coding, information, prediction, and estimation. In *IEEE Transactions on Information Theory*, volume 30, pages 629–636, 1984.

140. K. Rohr: Localization properties of direct corner detectors. In *Journal of Mathematical Imaging and Vision*, volume 4, pages 139–150, 1994.

141. F. Röhrdanz, H. Mosemann, and F. Wahl: HighLAP – A high level system for generating, representing, and evaluating assembly sequences. In *International Journal on Artificial Intelligence Tools*, volume 6, pages 149–163, 1997.

142. F. Röhrdanz, H. Mosemann, and F. Wahl: Geometrical and physical reasoning for automatic generation of grasp configurations. In *IASTED International Conference on Robotics and Manufacturing*, International Association of Science and Technology for Development, Anaheim, California, 1997.

143. C. Rothwell: Hierarchical object description using invariants. In J. Mundy, A. Zisserman, and D. Forsyth, editors, *Applications of Invariance in Computer Vision*, Lecture Notes in Computer Science, volume 825, pages 397–414, Springer Verlag, Berlin, 1993.

144. M. Salganicoff, L. Ungar, and R. Bajcsy: Active learning for vision-based robot grasping. In *Machine Learning*, volume 23, pages 251–278, 1996.

145. P. Sandon: Simulating visual attention. In *Journal of Cognitive Neuroscience*, volume 2, pages 213–231, 1990.

146. S. Sarkar and K. Boyer: Perceptual organization in Computer Vision – A review and a proposal for a classificatory structure. In *IEEE Transactions on Systems, Man, and Cybernetics*, volume 23, pages 382–399, 1993.

147. S. Sarkar and K. Boyer: A computational structure for preattentive perceptual organization – Graphical enumeration and voting methods. In *IEEE Transactions on Systems, Man, and Cybernetics*, volume 24, pages 246–267, 1994.

148. B. Schiele and J. Crowley: Object recognition using multidimensional receptive field histograms. In *European Conference on Computer Vision*, Lecture Notes in Computer Science, volume 1064, pages 610–619, Springer Verlag, Berlin, 1996.

149. R. Schalkoff: *Pattern Recognition – Statistical, Structural, and Neural Approaches*. John Wiley and Sons, New York, 1992.

150. A. Schmidt: *Manipulatorregelung durch fortwährende visuelle Rückkopplung*. Master thesis (in german), Christian-Albrechts-Universität zu Kiel, Institut für Informatik und Praktische Mathematik, Germany, 1998.

151. B. Schölkopf, K. Sung, and C. Burges: Comparing support vector machines with Gaussian kernels to radial basis function classifiers. In *IEEE Transactions on Signal Processing*, volume 45, pages 2758–2765, 1997.

152. B. Schölkopf: *Support Vector Learning*. PhD thesis, Technische Universität Berlin, Institut für Informatik, Germany, 1997.

153. G. Schöner, M. Dose, and C. Engels: Dynamics of behavior – Theory and applications for autonomous robot architectures. In *Robotics and Autonomous Systems*, volume 16, pages 213–245, 1995.

154. J. Segman, J. Rubinstein, and Y. Zeevi: The canonical coordinates method for pattern deformation – Theoretical and computational considerations. In *IEEE Transactions on Pattern Analysis and Machine Intelligence*, volume 92, pages 1171–1183, 1992.

155. N. Sharkey, editor: *Robotics and Autonomous Systems*. Special journal issue on Robot Learning – The New Wave, volume 22, pages 179–406, Elsevier Science Publishers, Amsterdam, The Netherlands, 1997.

156. K. Shimoga: Robot grasp synthesis algorithms – A survey. In *International Journal of Robotics Research*, volume 15, pages 230–266, 1996.

157. P. Simard, Y. LeCun, J. Denker, and B. Victorri: Transformation invariance in pattern recognition – Tangent distance and tangent propagation. In G. Orr and K.-P. Müller, editors, *Neural Networks – Tricks of the Trade*, Lecture Notes in Computer Science, volume 1524, pages 239–274, Springer Verlag, Berlin, 1998.

158. E. Simoncelli and H. Farid: Steerable wedge filters for local orientation analysis. In *IEEE Transactions on Image Processing*, volume 5, pages 1377–1382, 1996.

159. S. Smith and J. Brady: SUSAN – A new approach to low level image processing. In *International Journal of Computer Vision*, volume 23, pages 45–78, 1997.

160. A. Smola, B. Schölkopf, and K.-P. Müller: The connection between regularization operators and support vector kernels. In *Neural Networks*, volume 11, pages 637–649, 1998.

161. G. Sommer: Algebraic aspects of designing behavior based systems. In G. Sommer and J. Koenderink, editors, *Algebraic Frames for the Perception-Action Cycle*, Lecture Notes in Computer Science, volume 1315, pages 1–28, Springer Verlag, Berlin, 1997.

162. G. Sommer: The global algebraic frame of the perception-action cycle. In B. Jähne, editor, *Handbook of Computer Vision and Applications*, Academic Press, San Diego, California, pages 221–264, 1999.

163. K.-K. Sung and T. Poggio: *Example-based Learning for View-based Human Face Detection*. Artificial Intelligence Memo 1521, Massachusetts Institute of Technology, Cambridge, Massachusetts, 1994.

164. M. Swain and D. Ballard: Color indexing. In *International Journal of Computer Vision*, volume 7, pages 11–32, 1991.

165. M. Tipping and C. Bishop: Mixtures of probabilistic principal component analyzers. In *Neural Computation*, volume 11, pages 443–482, 1999.

166. J. Tsotsos: On the relative complexity of active versus passive visual search. In *International Journal of Computer Vision*, volume 7, pages 127–141, 1992.

167. J. Tsotsos: Behaviorist intelligence and the scaling problem. In *Artificial Intelligence*, volume 75, pages 135–160, 1995.

168. M. Turk and A. Pentland: Eigenfaces for recognition. In *Journal of Cognitive Neuroscience*, volume 3, pages 71–86, 1991.

169. P. Utgoff: *Machine Learning of Inductive Bias*. Kluwer Academic Publishers, Dordrecht, The Netherlands, 1986.

170. L. Valiant: A theory of the learnable. In *Communications of the ACM*, volume 27, pages 1134–1142, 1984.

171. V. Vapnik: *The Nature of Statistical Learning Theory*. Springer Verlag, Berlin, 1995.

172. F. Wahl and H. Biland: Decomposing of polyhedral scenes in Hough space. In *International Conference on Pattern Recognition*, pages 78–84, IEEE Computer Society Press, Los Alamitos, California, 1989.

173. C.-C. Wang: Extrinsic calibration of a vision sensor mounted on a robot. In *IEEE Transactions on Robotics and Automation*, volume 8, pages 161–175, 1992.

174. S. Waterhouse: *Classification and Regression using Mixtures of Experts*. PhD thesis, Jesus College, Cambridge, England, 1997.

175. H. Wechsler: *Computational Vision*. Academic Press, San Diego, California, 1990.

176. D. Williams and M. Shah: A fast algorithm for active contours and curvature estimation. In *Computer Vision and Image Understanding*, volume 55, pages 14–26, 1992.

177. P. Winston: *Artificial Intelligence*. Addison-Wesley Publishing Company, Reading, Massachussetts, 1992.

178. M. Yang, J.-S. Lee, C.-C. Lien, and C.-L. Huang: Hough transform modified by line connectivity and line thickness. In *IEEE Transactions on Pattern Analysis and Machine Intelligence*, volume 19, pages 905–910, 1997.

179. A. Ylä-Jääski and F. Ade: Grouping symmetric structures for object segmentation and description. In *Computer Vision and Image Understanding*, volume 63, pages 399–417, 1996.

180. Y. You, W. Xu, A. Tannenbaum, and M. Kaveh: Behavioral analysis of anisotropic diffusion in image processing. In *IEEE Transactions on Image Processing*, volume 5, pages 1539–1553, 1996.

181. W. Zangenmeister, H. Stiehl, and C. Freksa: *Visual Attention and Control*. Elsevier Science Publishers, Amsterdam, The Netherlands, 1996.

182. A. Zisserman, D. Forsyth, and J. Mundy: 3D object recognition using invariance. In *Artificial Intelligence*, volume 78, pages 239–288, 1995.

Lecture Notes in Computer Science

For information about Vols. 1–1964
please contact your bookseller or Springer-Verlag